AF847207

Biomedical Antenna Technology

Akinola Segun Ayokunle • Indrasen Singh •
Vivek Rajpoot • Ajay Kumar

Biomedical Antenna Technology

Shaping the Future of Healthcare

Springer

Akinola Segun Ayokunle
University of Johannesburg
Johannesburg, South Africa

Indrasen Singh
Vellore Institute of Technology
Vellore, Tamil Nadu, India

Vivek Rajpoot
Vellore Institute of Technology
Vellore, Tamil Nadu, India

Ajay Kumar
IMS Engineering College
Ghaziabad, Uttar Pradesh, India

ISBN 978-3-031-95267-8 ISBN 978-3-031-95268-5 (eBook)
https://doi.org/10.1007/978-3-031-95268-5

© The Editor(s) (if applicable) and The Author(s), under exclusive license to Springer Nature Switzerland AG 2026

This work is subject to copyright. All rights are solely and exclusively licensed by the Publisher, whether the whole or part of the material is concerned, specifically the rights of translation, reprinting, reuse of illustrations, recitation, broadcasting, reproduction on microfilms or in any other physical way, and transmission or information storage and retrieval, electronic adaptation, computer software, or by similar or dissimilar methodology now known or hereafter developed.
The use of general descriptive names, registered names, trademarks, service marks, etc. in this publication does not imply, even in the absence of a specific statement, that such names are exempt from the relevant protective laws and regulations and therefore free for general use.
The publisher, the authors and the editors are safe to assume that the advice and information in this book are believed to be true and accurate at the date of publication. Neither the publisher nor the authors or the editors give a warranty, expressed or implied, with respect to the material contained herein or for any errors or omissions that may have been made. The publisher remains neutral with regard to jurisdictional claims in published maps and institutional affiliations.

This Springer imprint is published by the registered company Springer Nature Switzerland AG
The registered company address is: Gewerbestrasse 11, 6330 Cham, Switzerland

If disposing of this product, please recycle the paper.

Preface

The rapid evolution of biomedical technologies over the past few decades has revolutionized the way we monitor and treat health conditions, providing significant advancements in the healthcare industry. One such area, which has gained tremendous attention, is the development and application of antenna technology in healthcare systems. The integration of antennas into medical devices, particularly in wearable and implantable health monitoring systems, has unlocked a new era of connectivity, data acquisition, and real-time medical intervention. It is with great enthusiasm that we present this book, *Biomedical Antenna Technology: Advancements for Healthcare Applications*, as a comprehensive exploration of the crucial role antennas play in advancing healthcare applications and the underlying technologies driving this transformation.

This book aims to provide a detailed and up-to-date overview of the design, development, and implementation of antenna technologies in biomedical applications. While antennas have been an essential component in various communications technologies for years, their application in healthcare, particularly in the realm of wearable devices, implantable sensors, medical telemetry, and telemedicine, is a rapidly expanding field that requires both technical understanding and careful consideration of unique constraints such as biocompatibility, miniaturization, energy efficiency, and user safety.

As a researcher and practitioner in the field, I have been fascinated by the potential of antenna technologies to enable advanced healthcare solutions. From monitoring vital signs remotely through wearable health devices to the ability to transmit real-time medical data from inside the human body through implantable sensors, antennas are crucial in creating seamless communication between healthcare devices and central monitoring systems. This ability to communicate data wirelessly has not only improved patient outcomes but has also paved the way for more efficient healthcare delivery, making it possible for medical professionals to make better-informed decisions quickly.

In the chapters of this book, we will explore various antenna designs and technologies that are pushing the boundaries of what is possible in the healthcare sector. The chapters are organized to introduce the reader to the different types of antennas used in medical applications, ranging from microstrip patch antennas for wearable health monitoring systems, to dipole antennas for medical telemetry, and helical antennas for capsule endoscopy. Each chapter offers a thorough discussion of the design considerations, challenges, innovations in these technologies, and the latest research developments in the field.

Chapter 1 begins with an introduction to the significance of antennas in healthcare applications, providing readers with a broad overview of the unique challenges and considerations that come with designing antennas for medical use. The chapter sets the stage by discussing the performance metrics for antennas in healthcare devices, including the crucial aspects of wireless communication, biocompatibility, and interference mitigation. It also highlights advancements in antenna technology and their impact on improving healthcare solutions, concluding with an outline of the book's structure and what the reader can expect from the following chapters.

Chapter 2 focuses on microstrip patch antennas, which are widely used in wearable health monitoring systems due to their compact size, low profile, and ease of integration with flexible materials. This chapter investigates design considerations, antenna miniaturization techniques, and methods to enhance antenna efficiency and radiation pattern, while addressing challenges related to antenna detuning and interference from other wireless devices. It also provides a case study that illustrates the practical application of microstrip patch antennas in the design of fitness trackers.

Chapter 3 explores the use of dipole antennas in medical telemetry systems, highlighting their advantages, limitations, and the technical challenges involved in using them for medical applications. This chapter also discusses their use in implantable medical devices, frequency band selection, and strategies to enhance the signal-to-noise ratio in telemetry systems.

Chapter 4 investigates helical antennas, which are particularly suitable for capsule endoscopy and wireless body area networks (WBAN). The unique advantages of helical antennas, including their circular polarization and broadband performance, make them ideal for biomedical applications. This chapter provides a detailed look at miniaturization techniques, biocompatibility considerations, and antenna design validation for helical antennas in medical applications.

Chapter 5 introduces the concept of planar inverted-F antennas (PIFA) in wireless medical devices. The chapter covers the advantages and limitations of PIFAs, design considerations, and their integration in medical devices. Tuning techniques and related simulation results are also discussed, providing readers with a deeper understanding of how PIFAs can be optimized for specific medical applications.

Chapter 6 shifts focus to implantable antennas, exploring the challenges and considerations for designing antennas that can be embedded in the human body for continuous health monitoring. This chapter discusses biocompatible materials, frequency range selection, and simulation results, as well as the open research challenges in the development of implantable antennas.

Chapter 7 covers multiple-input multiple-output (MIMO) antennas, which are increasingly being used in telemedicine applications to enhance communication capabilities and improve signal quality. This chapter provides insights into MIMO antenna design, frequency-selective surfaces for gain enhancement, and techniques to reduce side lobe levels, along with related research and simulation results.

Chapter 8 is dedicated to ultra-wideband (UWB) antennas, which have become a key technology for medical imaging and sensing. The chapter discusses the principles of UWB antennas, their design considerations for imaging and sensing devices, and their role in advancing medical diagnostics.

Chapter 9 explores the role of reconfigurable antennas in adaptive healthcare devices. This chapter highlights the potential of reconfigurable antennas to optimize performance in dynamic medical environments, offering solutions for frequency range adaptation, device-specific design, and overcoming challenges in real-time applications.

Chapter 10 concludes the book by examining the future trends and emerging technologies in biomedical antenna applications. This chapter provides a forward-looking perspective on promising research directions, challenges to be addressed, and the vision for the next generation of healthcare antennas.

Throughout this book, we have attempted to strike a balance between technical depth and accessibility. Whether you are an academic, researcher, or practitioner in the field of biomedical engineering or someone with a general interest in healthcare technology, we hope this book provides valuable insights into the intersection of antenna technology and healthcare applications.

We are deeply grateful to all the researchers and practitioners whose work has contributed to the development of biomedical antenna technologies. Their relentless pursuit of knowledge, innovation, and excellence continues to inspire and shape the future of healthcare. As we stand at the cusp of new advancements in medical diagnostics, treatment, and patient care, the role of antenna technologies in enabling these innovations cannot be overstated.

We invite you to explore the chapters that follow and hope that they serve as a resource for further research, development, and application of antennas in the exciting and rapidly evolving field of biomedical healthcare.

The authors wish to express their sincere gratitude to the many colleagues who offered invaluable suggestions throughout the preparation of this manuscript. The first author extends heartfelt thanks to his parents for their immeasurable blessings and sacrifices, which have consistently inspired the writing of this manuscript and the growth of his career. Their unwavering support and patience, in countless forms, cannot be fully conveyed in words. Special thanks are also extended to his wife, sons, and daughters, whose joyful smiles and conversations bring much-needed cheer, even during the most challenging times.

The authors are grateful to the publishers at Springer, particularly the Springer staff responsible for production, for their guidance and encouragement throughout the creation of this book. We also acknowledge the vital support of our family members, whose strength has been a constant source of motivation, driving us to work relentlessly toward the completion of this project. Lastly, we offer our deepest gratitude to the Almighty, the ultimate source of energy in the universe, for providing us with the strength and vitality to bring this work to fruition. All praise and thanks are due to Him.

Johannesburg, South Africa — Akinola Segun Ayokunle
Vellore, Tamil Nadu, India — Indrasen Singh
Vellore, Tamil Nadu, India — Vivek Rajpoot
Ghaziabad, Uttar Pradesh, India — Ajay Kumar

Contents

Chapter 1
Importance of Antennas in Healthcare Applications

Contents

1.1 Introduction

Biomedical antenna technology has emerged as a cornerstone in the advancement of modern healthcare systems. As the demand for efficient, precise, and minimally invasive diagnostic and therapeutic tools continues to rise, the integration of antenna systems in medical devices is revolutionizing how healthcare services are delivered [1]. From wireless implants that monitor vital signs to antennas in imaging systems that enhance diagnostic accuracy, this field has paved the way for unprecedented innovations, enabling better health outcomes globally.

© The Author(s), under exclusive license to Springer Nature Switzerland AG 2026
A. Segun Ayokunle et al., *Biomedical Antenna Technology*,
https://doi.org/10.1007/978-3-031-95268-5_1

The development of antennas in healthcare can be traced back to the broader evolution of wireless communication technologies [2]. Just as the telecommunications industry transformed with advancements from the first-generation (1G) analog systems to the high-speed, low-latency fifth-generation (5G) networks, biomedical antennas have seen their own specialized trajectory. Early antennas were rudimentary in design, primarily supporting basic wireless communication [3]. However, as healthcare systems evolved to demand real-time data transfer, wearable sensors, and wireless implants, the focus shifted toward developing antennas with unique specifications tailored to biomedical applications.

One of the significant milestones in biomedical antenna technology is its integration with the Internet of Medical Things (IoMT). This synergy has enabled continuous, remote patient monitoring, facilitating proactive healthcare interventions [4]. Antennas, as critical components of IoMT devices, have enabled seamless communication between patients, healthcare providers, and centralized data systems, ensuring real-time feedback and personalized care [5].

In diagnostic systems, antennas have significantly enhanced imaging modalities like magnetic resonance imaging (MRI), microwave imaging, and hyperthermia treatments [6]. High-frequency antennas used in these applications provide exceptional resolution, enabling early detection of diseases such as cancer. Similarly, in therapeutic interventions, antennas have demonstrated remarkable potential in targeted drug delivery, thermal ablation, and bio-stimulation, offering minimally invasive alternatives to traditional procedures [7].

The unique requirements of biomedical applications necessitate specialized antenna designs. Unlike conventional antennas used in telecommunications, biomedical antennas must function efficiently in complex and dynamic environments like the human body [8]. Factors like body tissue absorption, impedance matching, and miniaturization present significant design challenges. Additionally, ensuring the safety and biocompatibility of antennas is paramount, given their direct interaction with biological tissues.

Despite these challenges, technological advancements have led to the development of antennas capable of operating across a wide range of frequencies, including ultra-wideband (UWB), millimeter-wave (mmWave), and terahertz (THz) spectrums [9]. These advancements have unlocked new possibilities, such as real-time health monitoring, advanced diagnostic capabilities, and precision therapies, pushing the boundaries of what is achievable in modern medicine.

As the global healthcare landscape continues to face challenges such as aging populations, rising healthcare costs, and the increasing burden of chronic diseases, the need for innovative solutions becomes more pressing. Biomedical antennas, with their ability to enable non-invasive, efficient, and cost-effective healthcare delivery, are uniquely positioned to address these challenges [10]. By facilitating early disease detection, personalized treatment, and remote monitoring, these technologies are not only improving patient outcomes but also contributing to the sustainability of healthcare systems.

This book investigates into the fascinating world of biomedical antenna technology, exploring its theoretical foundations, design principles, and practical applications. It begins by tracing the historical development of antenna technology and its adaptation to meet biomedical needs. Subsequent chapters explore cutting-edge research and advancements in antenna design, material innovations, and their integration into medical devices. Case studies highlight real-world applications, demonstrating how these technologies are transforming healthcare practices globally.

By shedding light on the intricate interplay between biomedical engineering and wireless communication, this book aims to inspire researchers, engineers, and healthcare professionals to explore new frontiers in antenna technology. It underscores the pivotal role of interdisciplinary collaboration in addressing complex healthcare challenges and envisions a future where biomedical antennas become indispensable tools in the quest for better health outcomes.

1.2 Overview of Biomedical Antenna Design Challenges

Biomedical antenna design is a highly specialized field requiring engineers to balance performance, size, and functionality while meeting stringent safety and biocompatibility standards [11]. Designing antennas for healthcare applications is uniquely challenging because they must operate efficiently in dynamic environments, often within or near the human body. Factors such as tissue absorption, electromagnetic interference, miniaturization, and energy efficiency all present hurdles that must be addressed to create reliable and effective biomedical antennas. This section provides an in-depth discussion of these challenges and the innovative solutions driving advancements in this field.

1.2.1 Tissue Absorption and Electromagnetic Compatibility

One of the primary challenges in biomedical antenna design is the high degree of signal attenuation caused by the human body [12]. Human tissues, composed largely of water and electrolytes, exhibit significant dielectric properties that absorb electromagnetic waves. This leads to reduced antenna efficiency and poses a challenge for maintaining robust communication. Moreover, tissue absorption increases the risk of localized heating, which must be minimized to meet safety standards outlined by organizations such as the Federal Communications Commission (FCC) and International Commission on Non-Ionizing Radiation Protection (ICNIRP) [13].

To address this, designers often employ advanced simulation tools to model human tissue interactions and optimize antenna designs. For example, techniques such as impedance matching and the use of low-loss materials are employed to improve efficiency [14]. Additionally, the integration of frequency-selective surfaces and metamaterials has proven effective in enhancing electromagnetic compatibility by reducing interference and ensuring compliance with regulatory standards.

1.2.2 Miniaturization and Integration with Medical Devices

Modern healthcare demands devices that are small, lightweight, and unobtrusive. Implantable and wearable antennas must be miniaturized without compromising their performance [15]. This is particularly important for applications such as cardiac monitoring, glucose sensing, and drug delivery systems, where the antenna is integrated into compact medical devices.

Miniaturization poses unique challenges due to the trade-off between size and performance. Antennas must maintain adequate gain, bandwidth, and efficiency despite their reduced dimensions [16]. Techniques such as the use of high-permittivity substrates, meandered structures, and fractal geometries have emerged as solutions to achieve this balance. These designs allow for compact form factors while maintaining electromagnetic performance.

Integration with medical devices further complicates the design process. Antennas must be seamlessly incorporated into devices without causing mechanical or electrical interference. Flexible and stretchable materials, such as conductive polymers and liquid metals, are increasingly being used to enable conformal designs that adapt to the shape and movement of the human body. Advances in 3D printing technology have also opened new possibilities for fabricating intricate antenna structures tailored to specific medical applications [17].

1.2.3 Energy Efficiency and Power Management

Energy efficiency is a critical consideration for biomedical antennas, particularly for implantable devices that rely on limited power sources such as batteries or energy harvesting systems [18]. These devices must operate reliably for extended periods, often without the possibility of battery replacement. To enhance energy efficiency, researchers focus on optimizing antenna impedance matching, reducing power consumption, and improving radiation efficiency (see Fig. 1.1). These advancements enhance the energy efficiency and longevity of implantable biomedical devices. Wireless power transfer (WPT) and energy harvesting technologies, such as radio frequency (RF) energy harvesting and piezoelectric energy generation, are gaining traction as potential solutions. These approaches not only extend the operational life of biomedical devices but also eliminate the need for frequent invasive procedures to replace batteries.

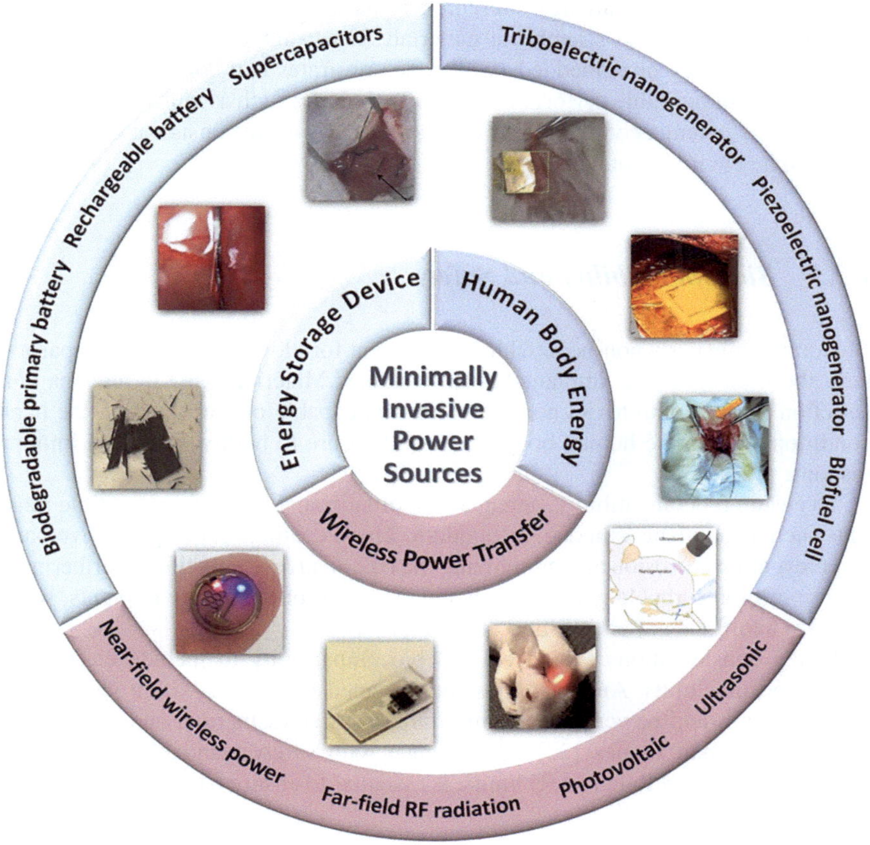

Fig. 1.1 Illustration of various alternative minimally invasive power sources for biomedical devices. These include biodegradable primary batteries, rechargeable batteries, and supercapacitors for energy storage. Additionally, triboelectric and piezoelectric nanogenerators harvest mechanical energy, while biofuel cells generate power through biological reactions. Wireless power transfer methods such as ultrasonic, photovoltaic, far-field RF, and near-field RF wireless power transfer enable continuous energy supply without the need for frequent battery replacements [19]

1.2.4 Dynamic Environments and Wearable Antennas

Biomedical antennas must perform consistently in dynamic environments where factors such as body movement, posture changes, and varying tissue properties can affect their performance [20]. Wearable antennas, in particular, face challenges in maintaining reliable connectivity due to the constant motion and deformation caused by the human body.

To mitigate these challenges, researchers have developed robust designs that incorporate flexible and stretchable materials. For example, antennas made from conductive textiles or embedded in silicone substrates can conform to the body while maintaining stable performance [21]. Techniques such as multi-layer stacking and diversity antenna configurations are also used to ensure consistent communication despite environmental changes.

1.2.5 Biocompatibility and Safety

Biocompatibility is a non-negotiable requirement for biomedical antennas, particularly those designed for implantable applications. Materials used in antenna construction must be non-toxic, non-irritating, and capable of withstanding the harsh conditions within the human body, such as exposure to bodily fluids and immune responses [22].

To ensure biocompatibility, antennas are often coated with biocompatible materials such as parylene, silicone, or titanium. These coatings act as protective barriers, preventing corrosion and minimizing the risk of adverse reactions. Furthermore, the use of bioresorbable materials is being explored for temporary implants, which dissolve harmlessly after fulfilling their purpose.

Safety considerations extend beyond biocompatibility to include electromagnetic exposure limits. Antennas must be designed to operate within specific frequency ranges and power levels to minimize the risk of thermal and non-thermal effects on human tissues [23]. Compliance with safety standards, such as the Specific Absorption Rate (SAR) limits, is essential to ensure patient safety.

1.2.6 Advancements in Materials and Fabrication Techniques

The development of advanced materials and fabrication techniques has played a pivotal role in overcoming the challenges of biomedical antenna design [24]. High-performance materials such as liquid crystal polymers (LCPs), ceramics, and graphene offer exceptional electrical and mechanical properties, making them ideal for biomedical applications. These materials enable the creation of antennas with enhanced efficiency, flexibility, and durability.

Innovative fabrication techniques, including inkjet printing, laser machining, and microfabrication, have further expanded the possibilities for biomedical antenna design [25]. These methods allow for the precise fabrication of intricate structures at micro and nanoscale levels, enabling the integration of antennas into compact and complex medical devices.

1.2.7 Emerging Solutions and Future Directions

The field of biomedical antenna design continues to evolve, driven by advancements in technology and the growing demand for innovative healthcare solutions. Emerging trends include the development of hybrid antennas that combine multiple functionalities, such as communication, sensing, and energy harvesting [26]. These multifunctional antennas are poised to revolutionize applications such as remote patient monitoring and telemedicine.

Artificial intelligence (AI) and machine learning (ML) are also being leveraged to optimize antenna designs. By analyzing large datasets and simulating various scenarios, AI algorithms can identify optimal design parameters and predict performance outcomes with high accuracy. This approach accelerates the design process and enables the creation of antennas that meet the stringent requirements of biomedical applications.

Biomedical antenna design is a complex and multidisciplinary field that addresses critical challenges to enable advanced healthcare technologies [27]. From overcoming tissue absorption and ensuring electromagnetic compatibility to achieving miniaturization, energy efficiency, and biocompatibility, the journey of designing biomedical antennas is both demanding and rewarding. Through innovative materials, fabrication techniques, and enabling technologies, researchers are pushing the boundaries of what is possible, paving the way for a new era of healthcare innovation. As the field continues to advance, the integration of biomedical antennas into medical devices will undoubtedly play a transformative role in improving patient outcomes and enhancing the quality of care.

1.2.8 Antenna Performance Metrics for Healthcare Devices

The performance of a biomedical antenna is evaluated based on specific metrics tailored to healthcare applications. These metrics are critical in ensuring the reliability and functionality of the antenna within the constraints of its operational environment [28]. Some key performance metrics include:

1. **Radiation Efficiency**: Biomedical antennas must exhibit high radiation efficiency, even in environments where human tissues absorb significant electromagnetic energy. This is crucial for minimizing power loss and ensuring effective data transmission.
2. **Impedance Matching**: Proper impedance matching is vital to ensure maximum power transfer between the antenna and the connected circuitry. In biomedical devices, achieving this balance is more challenging due to the complex impedance of biological tissues.
3. **Bandwidth**: A broad bandwidth is essential for supporting high data rate applications, such as real-time health monitoring and imaging. Biomedical antennas

often operate within ultra-wideband (UWB) and millimeter-wave (mmWave) frequencies to meet these demands.
4. **Specific Absorption Rate (SAR)**: SAR measures the rate at which body tissues absorb electromagnetic energy. Keeping SAR levels within regulatory limits ensures the antenna's safety and minimizes tissue heating.
5. **Miniaturization**: Compact antennas are essential for wearable and implantable devices. Engineers must optimize designs to achieve high performance in small form factors without compromising functionality.
6. **Directional Gain and Coverage**: Biomedical antennas must provide adequate gain and coverage in challenging environments, such as inside the human body, where signal attenuation is significant.

By meeting these metrics, biomedical antennas can deliver reliable, efficient, and safe performance in various healthcare applications.

1.2.9 Enabling Technologies

The development of biomedical antennas has been significantly driven by advancements in enabling technologies. These innovations provide the foundation for designing high-performance antennas that meet the unique requirements of healthcare applications. Key enabling technologies include:

1.2.9.1 Material Science

Material science is at the forefront of biomedical antenna development. The choice of materials plays a critical role in antenna design, impacting factors such as conductivity, flexibility, and biocompatibility [29]. Conductive polymers, for instance, offer lightweight and flexible solutions for wearable antennas, enabling seamless integration with the human body. Biocompatible metals such as titanium and gold are commonly used in implantable antennas due to their corrosion resistance and compatibility with bodily tissues [30]. Flexible substrates like polydimethylsiloxane (PDMS) allow for the creation of antennas that can conform to complex body shapes while maintaining durability and performance. Advances in material science continue to unlock new possibilities for designing antennas that are safe, reliable, and efficient.

1.2.9.2 Metamaterials and Meta-Surfaces

Metamaterials, with their engineered electromagnetic properties, have revolutionized antenna design. These materials can manipulate electromagnetic waves in ways that traditional materials cannot, allowing for enhanced performance in biomedical applications. For example, metamaterials enable miniaturization by achieving the desired resonant frequency in smaller structures [31]. They also improve gain and reduce electromagnetic interference, making them ideal for implantable antennas that operate within the human body's lossy environment. Meta-surfaces, a two-dimensional variant of metamaterials, further enhance antenna performance by providing precise control over wave propagation. These innovations are critical for addressing the unique challenges posed by biomedical applications.

1.2.9.3 Micromachining and Nanotechnology

Micromachining and nanotechnology have enabled the precise fabrication of miniature antennas, which are essential for modern healthcare devices. Microelectromechanical systems (MEMS) technology allows for the creation of highly intricate antenna structures with minimal size and weight. These techniques are particularly valuable for implantable devices that require compact and lightweight designs. Nanotechnology, on the other hand, facilitates the manipulation of materials at the atomic level, enabling the development of antennas with enhanced electrical properties [32]. For instance, carbon-based nanomaterials like graphene and carbon nanotubes are being explored for their exceptional conductivity and flexibility, offering new opportunities for biomedical antenna design.

1.2.9.4 Simulation Tools

Advanced simulation tools play a crucial role in the design and optimization of biomedical antennas. Techniques such as the finite element method (FEM) and the method of moments (MoM) allow engineers to model antenna performance under realistic conditions, including interactions with human tissues. These tools provide valuable insights into parameters such as radiation patterns, impedance matching, and specific absorption rates (SAR) [33]. By simulating various scenarios, engineers can identify potential design flaws and make necessary adjustments before physical prototyping. This approach not only accelerates the design process but also ensures that the final product meets stringent performance and safety standards.

1.2.9.5 Additive Manufacturing

Additive manufacturing, commonly known as 3D printing, has opened new possibilities for fabricating customized biomedical antennas. This technology creates intricate geometries that are difficult or impossible to achieve with traditional manufacturing methods [34]. Additive manufacturing is particularly beneficial for wearable devices, as it enables the production of antennas tailored to individual users. For instance, 3D-printed antennas can be designed to conform precisely to the contours of a patient's body, ensuring optimal performance and comfort. Additionally, the use of biocompatible printing materials ensures that these antennas are safe for prolonged use in medical applications.

1.2.9.6 Wireless Power Transfer and Energy Harvesting Technologies

Another significant enabler of biomedical antenna innovation is the integration of wireless power transfer (WPT) and energy harvesting technologies. These technologies address the challenge of powering implantable devices, which often have limited battery life. WPT systems use electromagnetic fields to transfer energy from an external source to the device, eliminating the need for battery replacements [35]. Similarly, energy harvesting systems convert ambient energy, such as body heat or motion, into electrical power. These approaches not only extend the operational life of biomedical devices but also enhance their overall reliability and functionality.

1.2.9.7 Artificial Intelligence and Machine Learning

Artificial intelligence (AI) and machine learning (ML) are increasingly being leveraged to optimize the design and functionality of biomedical antennas. By analyzing large datasets and simulating various scenarios, AI algorithms can identify optimal design parameters and predict performance outcomes with high accuracy [36]. For instance, AI-driven tools can optimize antenna placement, impedance matching, and radiation efficiency in complex medical environments. This approach accelerates the development process and ensures that biomedical antennas meet the unique demands of healthcare applications.

1.2.9.8 Flexible and Stretchable Electronics

The advent of flexible and stretchable electronics has significantly impacted the design of wearable biomedical antennas. These technologies enable the creation of antennas that can adapt to the dynamic movements of the human body without compromising performance. Materials such as conductive textiles, liquid metals, and stretchable polymers are commonly used in these designs [37]. Flexible antennas are particularly valuable for applications like remote patient monitoring, where consistent performance is essential despite body movement and environmental changes. Stretchable designs further enhance user comfort and device durability, making them ideal for long-term use in healthcare settings.

Enabling technologies have been instrumental in advancing the field of biomedical antenna design. From material science and metamaterials to additive manufacturing and artificial intelligence, these innovations provide the tools and techniques necessary to overcome the challenges of designing antennas for healthcare applications. As these technologies continue to evolve, they will play an increasingly critical role in shaping the future of biomedical antennas, paving the way for more effective and reliable healthcare solutions.

1.2.10 Advancements for Healthcare Applications

Biomedical antenna technology has witnessed remarkable advancements, enabling innovative applications that were once considered science fiction. These advancements have expanded the scope of antennas in healthcare, from diagnostics to therapy and beyond. Below is an in-depth discussion of some notable breakthroughs:

1.2.10.1 Wireless Implantable Devices

Advances in antenna miniaturization and biocompatibility have paved the way for the development of wireless implantable devices. These devices are designed to monitor and treat conditions such as cardiovascular diseases, diabetes, and neurological disorders. For example, implantable cardiac devices equipped with antennas can wirelessly transmit real-time data about a patient's heart condition, enabling early detection of abnormalities [38]. Neurological implants are also leveraging antenna technology to stimulate specific regions of the brain, offering new hope for patients with Parkinson's disease and epilepsy. The ability to wirelessly communicate with these devices not only enhances their functionality but also reduces the risks associated with invasive procedures.

Wearable devices have become an integral part of modern healthcare, thanks to the integration of advanced antennas. These devices can continuously monitor vital signs such as heart rate, blood pressure, and oxygen saturation, providing real-time feedback to users and healthcare providers. Antennas in wearable monitors ensure reliable data transmission even in dynamic environments, such as during physical activity. Smartwatches and fitness trackers are typical examples, but the technology has expanded to include more specialized devices for managing chronic conditions [39]. For instance, wearable glucose monitors use antennas to communicate glucose levels to a paired smartphone app, enabling better diabetes management.

High-frequency antennas have revolutionized medical imaging technologies, offering improved resolution and diagnostic accuracy. For instance, Magnetic Resonance Imaging (MRI) relies on specialized antennas to generate and receive electromagnetic signals that produce detailed images of internal organs and tissues [40]. Microwave imaging and terahertz imaging are emerging technologies that use antennas to detect abnormalities such as tumors and lesions. These imaging modalities offer non-invasive diagnostic solutions, reducing the need for surgical biopsies. Moreover, advancements in antenna design have enabled portable imaging systems, making diagnostic tools more accessible in remote and underserved areas.

1.2.10.2 Targeted Drug Delivery

Antennas integrated into therapeutic devices have opened new possibilities for targeted drug delivery. These systems use wireless communication to precisely direct drugs to specific areas of the body, minimizing side effects and enhancing treatment efficacy. For example, antennas in implantable drug delivery systems can be programmed to release medication in response to specific physiological triggers, such as a rise in blood sugar levels [41]. This approach improves patient compliance and ensures that drugs are delivered precisely where needed, reducing the risk of systemic side effects.

The COVID-19 pandemic accelerated the adoption of telemedicine, highlighting the critical role of antennas in enabling reliable wireless communication between patients and healthcare providers. Antennas in telemedicine devices facilitate high-speed data transfer for video consultations, remote diagnostics, and real-time monitoring of patients. Wearable and implantable devices equipped with antennas can transmit vital health data to healthcare professionals, enabling timely interventions [42]. This technology is particularly beneficial for managing chronic conditions and providing care to patients in rural or underserved areas, where access to healthcare facilities may be limited. Advances in smart implants equipped with antennas have transformed the way physiological parameters are monitored. These implants can continuously collect data on parameters such as blood pressure, glucose levels, and neural activity, transmitting this information in real time to healthcare systems. Biosensors integrated with antennas offer similar capabilities, providing non-invasive solutions for monitoring health metrics. For example, smart dental implants can detect early signs of oral diseases and alert the patient or dentist through a connected app [43]. The real-time data provided by these devices enables proactive healthcare management, reducing the risk of complications.

1.2.10.3 Antennas in Personalized Medicine

Personalized medicine aims to tailor healthcare interventions to individual patients based on their unique genetic, physiological, and lifestyle factors. Biomedical antennas play a crucial role in this paradigm by enabling precise data collection and targeted treatments. Wearable devices equipped with antennas can monitor specific biomarkers, providing insights into a patient's health status [44]. This information can be used to customize treatment plans, ensuring that therapies are optimized for each individual. Additionally, implantable antennas can facilitate the delivery of personalized therapies, such as gene editing or regenerative medicine, by targeting specific cells or tissues.

Biomedical antennas have also found applications in elderly care and rehabilitation. Wearable devices equipped with antennas can monitor the health of elderly individuals, alerting caregivers in case of emergencies such as falls or abnormal vital signs. These devices can also track physical activity and provide feedback to support rehabilitation programs. For instance, antennas in exoskeletons and prosthetic devices enable real-time communication with control systems, enhancing their functionality and user experience [45]. This technology not only improves the quality of life for elderly individuals but also reduces the burden on healthcare systems.

Wireless capsule endoscopy is a breakthrough diagnostic tool enabled by biomedical antennas. These capsules, equipped with miniature antennas, can travel through the gastrointestinal tract, capturing high-resolution images and transmitting them wirelessly to an external receiver. This non-invasive procedure provides a comprehensive view of the digestive system, aiding in the diagnosis of conditions such as ulcers, Crohn's disease, and gastrointestinal bleeding. Advances in antenna design have improved the reliability and data transmission capabilities of these capsules, making them a valuable addition to modern diagnostic techniques.

Energy efficiency is critical for biomedical antennas, particularly in implantable and wearable devices that rely on battery power. Advances in energy-efficient antenna systems have extended the operational life of these devices, reducing the need for frequent recharging or battery replacement. Techniques such as energy harvesting and wireless power transfer have further enhanced the sustainability of biomedical devices [46]. For example, antennas can capture ambient energy from radiofrequency signals or body heat, converting it into usable power for the device. These innovations not only improve device longevity but also enhance patient convenience.

The Internet of Medical Things (IoMT) has emerged as a transformative trend in healthcare, and biomedical antennas are at the heart of this ecosystem [47]. IoMT refers to the network of connected medical devices that collect, analyze, and transmit health data. Antennas enable seamless communication between these devices, ensuring that data is transmitted securely and efficiently. For instance, wearable monitors equipped with antennas can send real-time health data to cloud-based platforms, which can be analyzed using artificial intelligence. This integration enhances the accuracy of diagnoses, streamlines healthcare workflows, and empowers patients to take an active role in managing their health.

The advancements in biomedical antenna technology have significantly expanded their applications in healthcare, offering innovative solutions for diagnostics, therapy, and patient monitoring. From wireless implantable devices and wearable health monitors to smart implants and telemedicine, these technologies are transforming the way healthcare is delivered [48]. As research and development continue to push the boundaries of antenna design, the future holds even greater promise for addressing global healthcare challenges and improving patient outcomes.

1.2.11 Features of Antenna Wireless Communication

Wireless communication systems form the backbone of modern biomedical devices, with antennas playing a critical role as the primary interface for transmitting and receiving signals. The distinctive features of antenna wireless communication make them indispensable in healthcare applications, enabling seamless and efficient operation. Below is an in-depth exploration of these features and their contributions to biomedical advancements:

1.2.11.1 Low-Power Consumption

Biomedical antennas are designed with a focus on energy efficiency, operating with minimal power to extend the battery life of wearable and implantable devices [49]. This feature is crucial for ensuring the longevity and usability of medical devices, particularly in applications where frequent recharging or battery replacement is impractical. For instance, implantable cardiac monitors rely on low-power antennas to transmit data over extended periods, reducing the need for invasive procedures to replace power sources. Advances in power management techniques, such as duty cycling and energy harvesting, have further enhanced the energy efficiency of biomedical antennas, making them more sustainable and patient-friendly.

The ability to operate across multiple frequency bands is a hallmark feature of modern biomedical antennas. Multi-frequency operation enables antennas to support diverse healthcare applications, ranging from ultra-wideband (UWB) imaging systems requiring high data rates to low-frequency bands suited for long-range communication. For example, UWB antennas are used in microwave imaging to detect tumors, while low-frequency antennas facilitate remote patient monitoring in telemedicine [50]. This versatility ensures that biomedical antennas can cater to the unique requirements of various medical devices and environments, enhancing their functionality and adaptability.

Biomedical antennas are often deployed in environments with high levels of electromagnetic interference (EMI), such as hospitals and clinical settings. Advanced antenna designs incorporate techniques to mitigate EMI, ensuring reliable operation and data transmission. For instance, the use of shielding materials and filtering mechanisms helps minimize interference from external sources, such as medical imaging equipment and wireless networks. Additionally, directional antennas and beamforming technologies enable focused signal transmission, reducing the impact of interference and enhancing communication reliability. These advancements ensure that biomedical devices perform consistently even in challenging electromagnetic environments.

1.2.11.2 Miniaturization and Flexibility

The miniaturization and flexibility of biomedical antennas are critical features that enable their integration into wearable fabrics and implantable devices. Miniaturized antennas are designed to occupy minimal space, making them ideal for applications where size and weight are constraints. For instance, antennas used in wireless capsule endoscopy are small enough to fit within a swallowable capsule, enabling non-invasive diagnostic procedures. Flexible antennas, on the other hand, are constructed using materials such as conductive polymers and flexible substrates, allowing them to conform to the contours of the human body. This flexibility enhances the comfort and usability of wearable devices, ensuring that they can be worn for extended periods without causing discomfort.

Security is a paramount concern in healthcare, where the transmission of sensitive patient data must comply with strict regulations such as the Health Insurance Portability and Accountability Act (HIPAA) [51]. Biomedical antennas are equipped with features that support secure communication, including encryption protocols and secure data transmission mechanisms. These features ensure that patient information remains confidential and protected from unauthorized access. For example, antennas in wearable health monitors use encrypted channels to transmit data to healthcare providers, safeguarding patient privacy. The integration of secure communication features in biomedical antennas not only enhances patient trust but also ensures compliance with regulatory standards.

Maintaining robust connectivity is essential for the reliable operation of biomedical devices, particularly in challenging environments such as within the human body. Biomedical antennas are designed to ensure stable and consistent data transmission, overcoming obstacles such as tissue absorption and signal attenuation. For example, antennas used in implantable devices are optimized to operate in the complex, lossy environment of human tissues, ensuring reliable communication with external receivers. Techniques such as impedance matching and adaptive tuning are employed to enhance signal strength and connectivity. These features enable biomedical devices to deliver accurate and timely information, supporting effective healthcare interventions.

Many healthcare applications require transmitting large volumes of data, such as high-resolution images from medical imaging systems or continuous monitoring data from wearable devices. Biomedical antennas are designed to support high data transmission rates, ensuring that critical information is delivered without delays [52]. For instance, antennas in microwave imaging systems transmit high-frequency signals to generate detailed diagnostic images, aiding in the early detection of diseases. The ability to transmit data quickly and efficiently is a key feature that enhances the performance and utility of biomedical devices.

1.2.11.3 Adaptability to Dynamic Environments

Biomedical antennas must operate reliably in dynamic environments, where factors such as body movement, changes in posture, and varying tissue properties can affect performance. Adaptive antenna designs address these challenges by dynamically adjusting their parameters to maintain optimal performance. For example, reconfigurable antennas can switch between frequency bands or modify their radiation patterns in response to changing conditions. This adaptability ensures that biomedical devices can deliver consistent results in real-world scenarios, enhancing their effectiveness and user experience.

The integration of biomedical antennas with smart systems, such as the Internet of Medical Things (IoMT), has expanded their capabilities and applications. Antennas serve as the communication interface between medical devices and IoMT platforms, enabling the seamless exchange of data. For example, wearable devices equipped with antennas can transmit real-time health data to cloud-based systems, which are analyzed using artificial intelligence algorithms. This integration supports remote monitoring, predictive analytics, and personalized healthcare, transforming the way medical care is delivered. The ability to interface with smart systems is a defining feature that enhances the functionality and impact of biomedical antennas.

Low latency is a critical requirement for applications where real-time data transmission is essential, such as telemedicine and remote surgery. Biomedical antennas are designed to minimize latency, ensuring that data is delivered with minimal delays. For instance, antennas in robotic surgical systems enable precise and timely communication between the surgeon and the robotic instruments, enhancing the safety and accuracy of the procedure. The emphasis on low-latency communication ensures that biomedical devices can meet the demands of time-sensitive healthcare applications.

Biomedical antennas are constructed using biocompatible materials to ensure their safety and compatibility with the human body. Materials such as titanium, gold, and biocompatible polymers are commonly used to minimize the risk of adverse reactions. This feature is particularly important for implantable devices, where prolonged contact with tissues and fluids requires non-toxic, corrosion-resistant, and durable materials. The use of biocompatible materials enhances the safety and reliability of biomedical antennas, supporting their widespread adoption in healthcare.

The features of antenna wireless communication play a pivotal role in the design and functionality of biomedical devices, enabling seamless data transmission, secure communication, and reliable operation in complex environments. From low-power consumption and multi-frequency operation to robust connectivity and integration with smart systems, these features ensure that biomedical antennas meet the stringent requirements of modern healthcare applications. As technology continues to evolve, the innovative features of biomedical antennas will drive further advancements in medical devices, transforming healthcare delivery and improving patient outcomes [53].

1.2.12 Biomedical Antenna Technology

Biomedical antenna technology represents the convergence of engineering, medicine, and material science to create innovative solutions that address pressing healthcare needs. This technology is defined by its ability to enable non-invasive, real-time, and highly accurate medical interventions [54]. Biomedical antennas are a cornerstone of modern healthcare, providing the wireless communication capabilities essential for diagnostic, monitoring, and therapeutic applications. This section investigates into the critical aspects of biomedical antenna technology, highlighting design innovations, integration with the Internet of Medical Things (IoMT), advancements in biocompatibility, therapeutic applications, and future prospects.

1.2.12.1 Design Innovations

The design of biomedical antennas has seen significant advancements aimed at meeting the unique requirements of healthcare applications. Engineers are developing novel antenna designs tailored for specific biomedical applications. Notable examples include:

1. **Fractal Antennas:** Fractal geometry has been employed to create antennas with multi-frequency operation and compact sizes. These antennas are particularly useful in wearable and implantable devices where space constraints are critical.
2. **Patch Antennas:** Widely used in wearable devices, patch antennas are lightweight, low-profile, and capable of operating at multiple frequencies. Their planar structure makes them ideal for integration into flexible substrates.
3. **Dielectric Resonator Antennas (DRAs):** DRAs offer high radiation efficiency and low loss, making them suitable for medical imaging and therapeutic applications. Their ability to function in complex environments, such as human tissues, is a key advantage.
4. **Flexible and Stretchable Antennas:** These antennas are fabricated using materials like conductive polymers and liquid metals, enabling seamless integration into wearable fabrics or even directly onto the skin. They maintain performance despite bending, stretching, or twisting.

These design innovations ensure that biomedical antennas meet the performance, size, and reliability demands of modern healthcare applications.

1.2.12.2 Integration with IoMT

The Internet of Medical Things (IoMT) has revolutionized healthcare by connecting devices to collect, analyze, and share data. Biomedical antennas serve as the backbone of IoMT, enabling wireless communication between interconnected devices. Key aspects of this integration include:

1. **Real-Time Monitoring:** Antennas enable continuous data transmission from wearable and implantable devices to healthcare systems, providing real-time insights into patients' health.
2. **Data Analytics and AI:** Antennas facilitate the collection of large volumes of data, which can be analyzed using AI to predict health trends, detect anomalies, and guide treatment decisions.

3. **Remote Healthcare:** IoMT-powered devices equipped with biomedical antennas enable telemedicine, allowing patients to receive care from the comfort of their homes. This is especially critical for managing chronic conditions.
4. **Device Interoperability:** Antennas ensure seamless communication between diverse IoMT devices, fostering an ecosystem where multiple devices work together to improve healthcare outcomes.

The integration of biomedical antennas with IoMT underscores their importance in creating smarter, more connected healthcare systems.

1.2.12.3 Advancements in Biocompatibility

Biocompatibility is a critical consideration in the development of biomedical antennas, especially for implantable devices. Materials research has focused on ensuring that antennas are safe for long-term use within the human body. Key advancements include:

1. **Biocompatible Metals:** Materials like titanium, gold, and platinum are commonly used in implantable antennas due to their non-reactive nature and compatibility with human tissues.
2. **Flexible Substrates:** Substrates like polydimethylsiloxane (PDMS) and liquid crystal polymers (LCP) are biocompatible and flexible, making them suitable for wearable and implantable applications.
3. **Encapsulation Techniques:** Antennas are often encapsulated in biocompatible coatings to prevent direct contact with tissues and minimize the risk of adverse reactions.
4. **Corrosion Resistance:** Advanced materials and coatings enhance the durability of antennas, ensuring they can withstand the harsh conditions within the human body, such as exposure to fluids and varying pH levels.

These advancements in biocompatibility have made it possible to develop safe and reliable antennas for long-term medical use.

1.2.12.4 Therapeutic Applications

Biomedical antennas are not limited to diagnostic and monitoring applications; they also play a critical role in therapies. Some of the notable therapeutic applications include:

1. **Hyperthermia Treatment for Cancer:** Antennas are used to deliver precise electromagnetic energy to tumor sites, raising their temperature and destroying cancer cells without harming surrounding healthy tissues [55].
2. **Targeted Drug Delivery:** Antenna systems integrated into therapeutic devices enable the controlled release of drugs to specific areas, reducing side effects and improving efficacy.

3. **Neurological Stimulation:** Implantable antennas are used in devices like deep brain stimulators and spinal cord stimulators to treat conditions such as Parkinson's disease and chronic pain.
4. **Regenerative Medicine:** Antennas facilitate therapies like tissue engineering by providing wireless power and communication to support cell growth and tissue regeneration.

These therapeutic applications demonstrate the versatility and impact of biomedical antenna technology in advancing medical treatments.

1.2.12.5 Future Prospects

The future of biomedical antenna technology is promising, with emerging trends and technologies poised to revolutionize the field. Some of the most exciting prospects include:

1. **6G Wireless Communication:** The advent of 6G networks will bring unprecedented data speeds, ultra-low latency, and enhanced connectivity, enabling more sophisticated biomedical devices and applications.
2. **Artificial Intelligence (AI):** AI-powered antennas will enable smarter communication systems capable of adapting to changing environments and optimizing performance in real time.
3. **Energy Harvesting:** Future antennas will incorporate energy harvesting capabilities, drawing power from ambient sources like body heat or electromagnetic waves, eliminating the need for batteries.
4. **Nanotechnology:** Advances in nanotechnology will enable the development of ultra-small antennas for nanoscale devices, opening up new possibilities in targeted therapies and diagnostics.
5. **Personalized Healthcare:** Antennas designed for personalized healthcare applications will cater to individual needs, providing tailored treatments and monitoring solutions.

The dynamic and rapidly evolving nature of biomedical antenna technology offers limitless possibilities for transforming healthcare delivery and improving quality of life. By leveraging emerging technologies, researchers and engineers can continue to push the boundaries of what is possible in this exciting field.

1.3 Organization of the Book

This book is structured to provide a comprehensive exploration of biomedical antenna technologies and their critical role in advancing healthcare applications. The organization is designed to progressively delve into various antenna types, their design considerations, and their applications in healthcare technologies, from wearable devices to implantable systems.

Chapter 1 introduces the significance of antennas in healthcare, emphasizing the importance of innovative designs and the technological advancements that are enhancing the functionality of biomedical devices. This chapter sets the stage for the discussions to follow by reviewing the challenges faced in biomedical antenna design and the enabling technologies that make such advancements possible.

Chapter 2 explores microstrip patch antennas used in wearable health monitoring systems. It covers design considerations, miniaturization techniques, and the integration of biocompatible materials. This chapter also addresses the challenges of antenna detuning, wireless communication with the human body, and methods to improve antenna efficiency.

In Chap. 3, the focus shifts to dipole antennas used in medical telemetry systems. This chapter examines the advantages, limitations, and radiation mechanisms of dipole antennas, their application in wearable devices, and the challenges of frequency selection, interference management, and noise reduction.

Chapter 4 discusses helical antennas, particularly in capsule endoscopy and wireless body area networks (WBANs). It highlights the characteristics of helical antennas, miniaturization techniques, and biocompatible designs, while also addressing challenges related to frequency selection and antenna performance validation.

Chapter 5 introduces planar inverted-F antennas (PIFA), discussing their integration in wireless medical devices. It covers the benefits of PIFAs, design considerations, tuning techniques, and their application in medical devices, as well as current challenges and research directions in PIFA technology.

In Chap. 6, the book examines implantable antennas used in medical devices. This chapter investigates the challenges of implantable antenna design, focusing on biocompatible materials, frequency range considerations, and the constraints of implantable devices in the human body.

Chapter 7 presents MIMO antennas for telemedicine applications. It explores the design and configuration of MIMO antennas, including frequency-specific considerations, hybrid-fed systems, and techniques for enhancing the antenna's gain. It also discusses the emerging role of MIMO antennas in telemedicine.

Chapter 8 focuses on ultra-wideband (UWB) antennas, which are used in medical imaging and sensing. This chapter covers the principles, design, and application of UWB antennas in healthcare, highlighting their potential for improving diagnostic and sensing capabilities in medical systems.

In Chap. 9, the concept of reconfigurable antennas is introduced, exploring their applications in adaptive healthcare devices. The chapter discusses frequency range considerations, antenna design, and the challenges of creating antennas that can dynamically adapt to changing healthcare needs.

Chapter 10 concludes the book by looking ahead to future trends and emerging antenna technologies in biomedical applications. This final chapter examines the current state of the field, promising advances, and research directions, envisioning the next generation of healthcare antennas.

1.4 Summary

The field of biomedical antenna technology has rapidly evolved, transforming healthcare delivery and improving patient outcomes. Modern biomedical antennas must address various challenges, including operating in dynamic environments, ensuring biocompatibility, and meeting performance metrics such as radiation efficiency, bandwidth, and specific absorption rate (SAR). Enabling technologies like advanced materials, metamaterials, micromachining, and simulation tools have played a pivotal role in overcoming these challenges, paving the way for innovative designs and applications.

Advancements in biomedical antennas have enabled groundbreaking applications, from wireless implantable devices to telemedicine and real-time health monitoring. These antennas integrate seamlessly with the Internet of Medical Things (IoMT), creating interconnected systems that deliver actionable insights for healthcare providers. Future prospects, such as leveraging 6G wireless communication, artificial intelligence, and ultra-wideband technologies, promise to revolutionize biomedical antenna technology further.

Biomedical antenna technology is positioned to address global healthcare challenges through non-invasive, precise, and highly reliable solutions. It will continue to advance through innovations in design, materials, and integration with emerging technologies, ultimately driving sustainable and effective healthcare systems.

References

1. L. Abdenasser, Miniaturization of Antennas and Wireless Power Transfer System for Biomedical Implantable Devices, 2024.
2. A. Se, A. Telukdarie, Revolutionizing healthcare delivery through wireless wearable antenna frameworks: Current trends and future prospects. IEEE Access **11**, 80327–80347 (2023)
3. A. Ghosh, *Wireless Communications* (Cambridge Univ. Press, Cambridge, U.K., 2005)
4. N. Kavitha, P. Palaniyammal, Advancements in biomedical antennas: exploring diverse designs for wireless connectivity in medical applications, in *Antennas for Industrial and Medical Applications with Optimization Techniques for Wireless Communication*, (CRC Press, Boca Raton, FL, USA, 2024), pp. 128–150
5. M. Basheer et al., Patient monitoring system based on internet of things: a review and related challenges with open research issues. IEEE Access (2024)
6. C. Ortega, D.O. Rodriguez-Duarte, J.A. Tobon Vasquez, J.-C. Bolomey, F. Vipiana, Review of microwave near-field sensing and imaging devices in medical applications. Sensors **24**(14), 4515 (2024)
7. S. Liu, Y. Rao, H. Jang, P. Tan, N. Lu, Strategies for body-conformable electronics. Matter **5**(4), 1104–1136 (2022)
8. S. Saravanan, K.U. Haimavathi, K. Aafizaa, Biomedical antenna design: microstrip patch antennas for wearable and implantable healthcare applications—A review. Biomed. Mater. Devices, 1–11 (2025)
9. K. Rashid, A. Iftikhar, B. Ijaz, I.B. Mabrouk, The potentials, challenges, and future directions of on-chip-antennas for emerging wireless applications—a comprehensive survey. IEEE Access **7**, 173897–173934 (2019)

10. V. Gupta, R. Kumar, Biomedical telemetry antenna innovations: progress, uses, and prospects for the future. Prog. Electromagn. Res. B **106** (2024)
11. J.M. Bangeles, E.M. Tapawan, D.E. Arboleda, Wearable antennas in medical applications: a comprehensive review of design, performance, safety, and clinical applications. Int. J. Res. Publ. Rev. **5**, 709–723 (2024)
12. K. Aliqab, I. Nadeem, S.R. Khan, A comprehensive review of in-body biomedical antennas: design, challenges and applications. Micromachines **14**(7), 1472 (2023)
13. L. Hardell, M. Nilsson, T. Koppel, M. Carlberg, Aspects on the international commission on non-ionizing radiation protection (ICNIRP) 2020 guidelines on radiofrequency radiation. J. Cancer Sci. Clin. Ther. **5**(2), 250–285 (2021)
14. A. Rice, A. Kiourti, High-contrast low-loss antenna: a novel antenna for efficient into-body radiation. IEEE Trans. Antennas Propag. **70**(11), 10132–10140 (2022)
15. Y. Zhu et al., Recent advancements and applications in 3D printing of functional optics. Addit. Manuf. **52**, 102682 (2022)
16. S. Roy et al., Powering solutions for biomedical sensors and implants inside the human body: a comprehensive review on energy harvesting units, energy storage, and wireless power transfer techniques. IEEE Trans. Power Electron. **37**(10), 12237–12263 (2022)
17. M. Xu et al., Minimally invasive power sources for implantable electronics. Exploration **4**(1), 20220106 (2024)
18. P.S. Hall, Y. Hao, *Antennas and Propagation for Body-Centric Wireless Communications, Norwood, MA* (Artech House, USA, 2012)
19. S.N. Mahmood et al., Recent advances in wearable antenna technologies: a review. Prog. Electromagn. Res. B **89**, 1–27 (2020)
20. H. Y. Atay, The fabrication and technical applications of multifunctional materials, Ph.D. dissertation, Dokuz Eylul Univ., Turkey, 2013.
21. J. da Silva Gomes, Therapeutic applications of electromagnetic radiation, Master's thesis, Univ. do Porto, Portugal, 2016.
22. S.M. Ali et al., Recent advances of wearable antennas in materials, fabrication methods, designs, and their applications: state-of-the-art. Micromachines **11**(10), 888 (2020)
23. S.G. Kirtania et al., Flexible antennas: a review. Micromachines **11**(9), 847 (2020)
24. M.A. Ullah et al., A review on antenna technologies for ambient RF energy harvesting and wireless power transfer: designs, challenges and applications. IEEE Access **10**, 17231–17267 (2022)
25. A. Alexiou, M. Haardt, Smart antenna technologies for future wireless systems: Trends and challenges. IEEE Commun. Mag. **42**(9), 90–97 (2004)
26. R.B. Green, M.V. Smith, E. Topsakal, In vitro and in vivo testing of implantable antennas, in *Antenna and Sensor Technologies in Modern Medical Applications*, (2021), pp. 145–189
27. G. Oliveri, D.H. Werner, A. Massa, Reconfigurable electromagnetics through metamaterials—a review. Proc. IEEE **103**(7), 1034–1056 (2015)
28. P. Tiwari, N. Sharma, The potential of nanoantennas: highlighting the impact of metamaterials integration, in *Handbook of Nano-Metamaterials*, (Springer Nature Singapore, Singapore, 2024), pp. 1–48
29. M. Abdul-Al et al., Wireless electromagnetic radiation assessment based on the specific absorption rate (SAR): a review case study. Electronics **11**(4), 511 (2022)
30. S.A.M. Tofail et al., Additive manufacturing: Scientific and technological challenges, market uptake and opportunities. Mater. Today **21**(1), 22–37 (2018)
31. J. Zhang et al., Battery-free and wireless technologies for cardiovascular implantable medical devices. Adv. Mater. Technol. **7**(6), 2101086 (2022)
32. K. Guo et al., Artificial intelligence and machine learning in design of mechanical materials. Mater. Horiz. **8**(4), 1153–1172 (2021)
33. N. Matsuhisa et al., Materials and structural designs of stretchable conductors. Chem. Soc. Rev. **48**(11), 2946–2967 (2019)
34. B.D. Nelson et al., Wireless technologies for implantable devices. Sensors **20**(16), 4604 (2020)

35. K. Bayoumy et al., Smart wearable devices in cardiovascular care: where we are and how to move forward. Nat. Rev. Cardiol. **18**(8), 581–599 (2021)
36. T. Herrmann et al., Metamaterial-based transmit and receive system for whole-body magnetic resonance imaging at ultra-high magnetic fields. PLoS One **13**(1), e0191719 (2018)
37. S.M. Mirvakili, R. Langer, Wireless on-demand drug delivery. Nat. Electron. **4**(7), 464–477 (2021)
38. A.R. Chishti et al., Advances in antenna-based techniques for detection and monitoring of critical chronic diseases: a comprehensive review. IEEE Access **11**, 104463–104484 (2023)
39. J. Kim et al., Wearable biosensors for healthcare monitoring. Nat. Biotechnol. **37**(4), 389–406 (2019)
40. T. He, C. Lee, Evolving flexible sensors, wearable and implantable technologies towards BodyNET for advanced healthcare and reinforced life quality. IEEE Open J. Circuits Syst. **2**, 702–720 (2021)
41. B. Bhushan et al., Towards a secure and sustainable internet of medical things (IoMT): Requirements, design challenges, security techniques, and future trends. Sustainability **15**(7), 6177 (2023)
42. G. Gaobotse, E. Mbunge, J. Batani, B. Muchemwa, Non-invasive smart implants in healthcare: redefining healthcare services delivery through sensors and emerging digital health technologies. Sensors Int. **3**, 100156 (2022)
43. Y. Alkhalaf, H.M.Y. Ahmad, H. Ramiah, Self-sustainable biomedical devices powered by RF energy: A review. Sensors **22**(17), 6371 (2022)
44. A.H. Dalloul, F. Miramirkhani, L. Kouhalvandi, A review of recent innovations in remote health monitoring. Micromachines **14**(12), 2157 (2023)
45. E. Isibor, Regulation of healthcare data security: legal obligations in a digital age. SSRN Electron J (2024) [Online]. Available: https://ssrn.com/abstract=4957244
46. H. Liu et al., Flexible wearable antenna based on graphene-printed textiles for body-centric communications. IEEE Access **9**, 85024–85032 (2021)
47. M. T. Islam, M. R. I. Faruque, and N. Misran, "Analysis of materials for wearable antenna in wireless body area network," Int J Antennas Propag, vol. 2012, Article ID 565821, 6 pages, 2012.
48. K. Agarwal, Y.X. Guo, N.V. Thakor, Wireless power transfer strategies for implantable bioelectronics. IEEE Rev Biomed Eng **10**, 136–161 (2017)
49. F. Merli et al., Design, realization and measurements of a miniature antenna for implantable wireless communication systems. IEEE Trans Antennas Propag **59**(10), 3544–3555 (2011)
50. A. Kiourti, K.S. Nikita, Miniature scalp-implantable antennas for telemetry in the MICS and ISM bands: Design, safety considerations and link budget analysis. IEEE Trans Antennas Propag **60**(8), 3568–3575 (2012)
51. A. Collado, A. Georgiadis, Conformal hybrid solar and electromagnetic (EM) energy harvesting rectenna. IEEE Trans Circuits Syst I Regul Pap **60**(8), 2225–2234 (2013)
52. T. Wu, T.S. Rappaport, C.M. Collins, Safe for generations to come: Considerations of safety for millimeter waves in wireless communications. IEEE Microw Mag **16**(2), 65–84 (2015)
53. M.S. Pathak, R. Kumar, Design and analysis of dual band wearable antenna for biomedical applications. Int. J. RF Microw. Comput. Aided Eng **30**(10), e22395 (2020)
54. N.S. Islam, S. Akter, M.T. Islam, Multiband compact wearable antenna for biomedical applications. Microw Opt Technol Lett **60**(9), 2263–2269 (2018)
55. S.P. Yadav, R. Kumar, M.S. Pathak, A review on flexible wearable antenna for biomedical applications. Mater. Today Proc. **62**, 4976–4980 (2022)

Chapter 2
Microstrip Patch Antennas for Wearable Health Monitoring Systems

Contents

© The Author(s), under exclusive license to Springer Nature Switzerland AG 2026

A. Segun Ayokunle et al., *Biomedical Antenna Technology*,
https://doi.org/10.1007/978-3-031-95268-5_2

2.1 Introduction

Wearable health monitoring devices are at the forefront of modern healthcare, providing continuous, real-time data about an individual's health and well-being [1]. These devices, ranging from fitness trackers and smartwatches to medical sensors and implantable devices, rely heavily on antennas for wireless communication. As the demand for more compact, efficient, and comfortable wearables increases, microstrip patch antennas (MPAs) have emerged as a popular solution due to their low-profile design, ease of integration, and suitable performance for the frequency bands used in health monitoring [2].

MPAs are a type of printed antenna that consists of a conducting patch on a dielectric substrate. They are well-suited for wireless communication in wearable health monitoring systems because of their small size, lightweight nature, and capability to operate at higher frequencies. These characteristics make MPAs particularly advantageous for applications where portability, comfort, and reliability are critical [3]. As wearables become increasingly integrated with advanced sensors and communication systems, understanding the principles of MPA design is crucial for ensuring optimal performance.

The rapid growth in wearable health monitoring systems has brought forward challenges such as optimizing antenna performance in proximity to the human body, ensuring energy efficiency, and maintaining reliable connectivity in diverse environments. Additionally, the miniaturization of wearable devices has necessitated the design of antennas that are not only efficient but also capable of coexisting with other components within compact form factors. This chapter investigates the various factors that influence the design and application of MPAs in wearable health monitoring devices. It covers the essentials of wireless communication systems and design considerations specific to wearable antennas and provides a case study demonstrating the design of an MPA for a fitness tracker.

2.2 Wireless Communication Systems

Wireless communication is the backbone of modern health monitoring systems, enabling real-time transmission of physiological data to remote healthcare providers or cloud-based systems for analysis [4]. Figure 2.1 illustrates the key components of a wireless communication transceiver, which include the transmitter (Tx) and receiver (Rx) sections. The input signal may consist of a spoken message, a video feed from a CCD camera, or an email transmission [5]. The integration of wireless communication technologies into health monitoring devices allows for continuous, non-invasive monitoring of a patient's vital signs, such as heart rate, blood pressure, and glucose levels, without physical connections. These devices are becoming increasingly sophisticated, utilizing various wireless standards such as Bluetooth, Zigbee, Wi-Fi, and cellular networks.

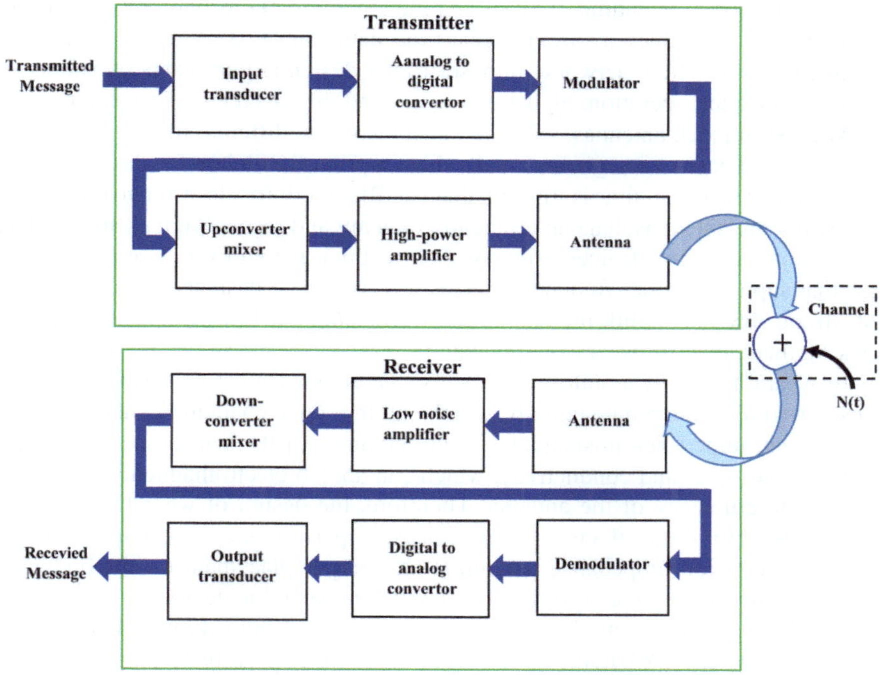

Fig. 2.1 The key components of a wireless communication transceiver [5]

The wireless communication system in wearable health monitors typically consists of three main components: the antenna, the radio frequency (RF) circuitry, and the data interface. The antenna plays a vital role in transmitting and receiving signals between the device and external systems. The choice of wireless communication technology and frequency band directly impacts the antenna design, as different frequencies have different propagation characteristics and power requirements.

For wearable devices, the most commonly used frequency bands fall within the Industrial, Scientific, and Medical (ISM) bands, which are unlicensed and allow for low-power communication over short to medium distances. The 2.4–5.8 GHz frequency range is particularly popular for wearable health monitors because it provides a good balance between range, power consumption, and data rate. Bluetooth, which operates in the 2.4 GHz ISM band, is a widely used standard in fitness trackers and medical devices due to its low-power consumption and ease of integration. Additionally, Zigbee, operating in the same frequency band, offers low-power consumption and is suitable for applications requiring mesh networking and long battery life.

More advanced wearable health devices, such as those supporting telemedicine applications, may utilize Wi-Fi or cellular networks for higher data rate transmissions and broader coverage. Wi-Fi, operating in the 2.4 GHz and 5 GHz bands, is ideal for high-speed data transfer but requires more power compared to Bluetooth.

Cellular networks, including 4G LTE and the emerging 5G technology, extend the capability of wearables to provide real-time data transmission across large distances. However, these networks impose stricter design requirements on antennas, such as wideband operation, higher efficiency, and improved radiation patterns.

Microstrip patch antennas, with their adaptability to different frequencies, compactness, and ease of integration, make them a prime candidate for these types of systems. For instance, fitness trackers that use Bluetooth for short-range communication require antennas that can efficiently transmit and receive signals at 2.4 GHz, while more advanced devices using Wi-Fi or cellular networks may require antennas that support broader frequency ranges. MPAs can be designed to cater to specific frequency bands while maintaining high efficiency and low-power consumption, making them ideal for wearable applications.

In wearable health monitoring systems, the antenna must perform reliably in close proximity to the human body, which can introduce challenges such as signal attenuation, impedance mismatching, and detuning [6]. Human tissue has a high dielectric constant and conductivity, which can absorb electromagnetic waves and reduce the efficiency of the antenna. Therefore, the design of wearable antennas must account for these effects to ensure consistent performance. Techniques such as substrate selection, impedance matching, and proper placement on the body are crucial for optimizing the performance of MPAs in wearable devices.

Another critical aspect of wireless communication systems in wearable devices is energy efficiency. Wearable health monitors often rely on battery-powered systems, making it essential to minimize power consumption without compromising performance. MPAs, with their inherently low profile and ability to operate efficiently at higher frequencies, are well suited for achieving this balance. By optimizing the antenna design, it is possible to enhance the overall energy efficiency of the device, thereby extending battery life and improving user experience.

The choice of substrate material also plays a significant role in the design and performance of MPAs for wearable health monitoring. Substrates with low dielectric constants are preferred for achieving wider bandwidth and improved radiation efficiency. Flexible substrates, such as polymers and textiles, are often used in wearable applications to enhance comfort and adaptability [7]. These materials allow MPAs to conform to the shape of the body, ensuring consistent performance even under dynamic conditions. The use of flexible and biocompatible materials further enables the integration of MPAs into clothing, wristbands, or directly onto the skin, expanding the possibilities for innovative wearable designs.

The integration of MPAs into wearable health monitoring systems also requires careful consideration of electromagnetic compatibility (EMC) and interference. Wearable devices often operate in environments with multiple wireless systems, such as smart homes, hospitals, and fitness centers ensuring that the MPA does not interfere with other wireless systems and vice versa is essential for maintaining reliable communication. Techniques such as frequency isolation, proper grounding, and shielding can help mitigate interference and improve the overall performance of the antenna.

Wireless communication systems form the foundation of modern wearable health monitoring devices, enabling seamless data transmission and real-time health tracking. Microstrip patch antennas, with their compact design, versatility, and adaptability to various frequency bands, play a pivotal role in these systems. By addressing challenges such as body proximity effects, energy efficiency, and electromagnetic compatibility, MPAs can be optimized to meet the stringent requirements of wearable health monitoring applications. The next section of this chapter will explore the specific design considerations of MPAs in wearable systems and provide a case study to illustrate their practical implementation in a fitness tracker.

2.3 Design Considerations for Wearable Antennas

Designing an antenna for a wearable device presents unique challenges compared to traditional antenna designs. Wearable devices are expected to be compact, efficient, lightweight, and user-friendly while providing consistent and reliable performance [8]. These requirements pose various design considerations that go beyond the standard parameters of antenna engineering. The human body introduces significant challenges, as it is both a source of interference and an absorber of electromagnetic waves. To address these challenges effectively, a comprehensive understanding of the following factors is essential.

First, size constraints play a pivotal role in the design of wearable antennas. Unlike traditional antennas, which can be relatively larger and mounted on fixed surfaces, wearable antennas must fit into compact devices such as smartwatches, fitness trackers, or medical patches [9]. This requires innovative approaches to miniaturization without compromising on performance. Microstrip patch antennas (MPAs) are particularly suitable for this purpose due to their low-profile structure and ability to operate efficiently at high frequencies. Achieving a balance between size and performance is one of the foremost challenges in wearable antenna design.

Second, efficiency is another critical factor. The antenna must be capable of transmitting and receiving signals effectively, even when subjected to environmental factors such as movement, bending, or contact with the skin [10]. Wearable devices are often in close proximity to the human body, which is a lossy dielectric medium that can significantly affect the antenna's performance. This proximity can lead to detuning, reduced efficiency, and signal attenuation. To mitigate these effects, antenna designs must incorporate measures such as impedance matching, high-gain configurations, and the use of protective materials to minimize interaction with the body.

Third, power consumption is a vital consideration. Most wearable devices rely on battery power, which imposes stringent requirements on energy efficiency. The antenna must be designed to operate at minimal power levels while maintaining reliable communication. This necessitates the use of advanced materials, optimized geometries, and integration with low-power electronics to extend battery life. The interplay between the antenna and the electronic components, such as RF transceivers and amplifiers, must be carefully calibrated to ensure optimal performance without unnecessary power drain.

Biocompatibility is a unique requirement for wearable antennas, especially those intended for long-term use or direct contact with the skin. Materials used in the antenna must be safe for human contact, non-irritating, and durable under various conditions, including sweat, temperature fluctuations, and mechanical stress. Additionally, the antenna must operate at power levels that do not pose any health risks to the user.

The following subsections delve deeper into specific aspects of wearable antenna design, focusing on electronic solid-state sources and the frequency range of 2.4–5.8 GHz.

2.3.1 Electronic Solid-State Sources

Electronic solid-state sources are the backbone of modern wireless communication systems, and their integration with antennas in wearable devices is no exception. These sources include components such as RF transceivers, power amplifiers, oscillators, and mixers, which are responsible for generating, modulating, and transmitting signals. The performance of these components directly influences the efficiency and reliability of the antenna system, making them a critical aspect of wearable antenna design.

In wearable health monitoring devices, the primary challenge is to achieve low-power consumption without compromising signal quality. Wearable devices are typically powered by small, lightweight batteries, which impose limitations on the amount of power available for wireless communication. To address this, solid-state sources must be designed to operate at minimal power levels while maintaining high performance. Advanced semiconductor technologies, such as complementary metal-oxide-semiconductor (CMOS) and gallium arsenide (GaAs), are commonly used to develop low-power RF components with excellent efficiency [11].

The integration of solid-state sources with microstrip patch antennas requires careful attention to impedance matching and signal integrity. Impedance matching ensures that the maximum amount of power is transferred between the source and the antenna, minimizing losses and improving overall efficiency. This is particularly important in wearable devices, where power conservation is a top priority. Engineers often use matching networks, such as inductors and capacitors, to achieve the desired impedance characteristics and optimize performance [12].

Another critical consideration is the interaction between the antenna and the solid-state sources in terms of frequency stability and noise reduction. Wearable devices often operate in environments with significant electromagnetic interference (EMI), which can degrade signal quality. Solid-state sources must be designed to provide stable and noise-free signals, even in challenging conditions. Techniques such as frequency hopping, error correction, and shielding are commonly employed to mitigate the effects of EMI and ensure reliable communication.

For example, Bluetooth transceivers, which are widely used in wearable health monitoring devices, are designed to operate at low-power levels while providing robust connectivity. These transceivers typically include features such as adaptive frequency hopping and power control to optimize performance in dynamic environments. When integrated with an MPA, the transceiver's performance is further enhanced by ensuring that the antenna's resonant frequency and bandwidth align with the Bluetooth protocol's requirements. This synergy between the solid-state source and the antenna is essential for achieving optimal performance in wearable devices [13].

In addition to Bluetooth, other wireless communication standards, such as Zigbee, Wi-Fi, and cellular networks, also rely on solid-state sources for signal generation and modulation. Each standard has its own unique requirements in terms of frequency, power, and data rate, necessitating tailored designs for the solid-state components and the antenna. The versatility of MPAs makes them well-suited for these applications, as they can be easily tuned to operate at the desired frequencies and provide the necessary performance metrics.

2.3.2 Frequency Range: 2.4–5.8 GHz

The frequency range of 2.4–5.8 GHz is a cornerstone of modern wireless communication, offering a balance between range, data rate, and power consumption. This frequency range encompasses the Industrial, Scientific, and Medical (ISM) bands, which are widely used for various applications, including wearable health monitoring devices [14]. The popularity of these bands can be attributed to their unlicensed status, making them accessible for low-power communication over short to medium distances.

The 2.4 GHz band is particularly well-suited for wearable devices due to its favorable propagation characteristics. Signals in this band can penetrate obstacles such as walls and human tissue, providing reliable connectivity even in challenging environments. Bluetooth and Zigbee are two prominent wireless standards that operate in the 2.4 GHz band, making it a popular choice for fitness trackers, smartwatches, and medical sensors. The compact size of MPAs at this frequency further enhances their suitability for wearable applications, where space is often limited.

The 5 GHz band, on the other hand, offers higher data rates and reduced interference compared to the 2.4 GHz band [15]. Wi-Fi is the primary standard operating in this band, providing seamless connectivity for more advanced wearable devices. The higher frequency of the 5 GHz band allows for smaller antenna sizes, which is advantageous for compact wearables. However, the propagation characteristics of this band are less favorable, as signals are more susceptible to attenuation and have a shorter range compared to the 2.4 GHz band. This trade-off must be carefully considered when designing wearable antennas for dual-band or multiband operation [16].

Designing MPAs for the 2.4–5.8 GHz frequency range involves several key considerations. The first is bandwidth, which determines the range of frequencies over which the antenna can operate effectively. Wearable devices often require broad bandwidths to accommodate multiple wireless standards and ensure reliable communication in dynamic environments. Engineers use techniques such as substrate selection, patch geometry optimization, and the addition of parasitic elements to enhance the bandwidth of MPAs [17].

Another important consideration is the antenna's radiation pattern, which defines the direction and strength of the emitted signal. For wearable devices, the radiation pattern must be designed to ensure omnidirectional coverage, allowing for reliable communication regardless of the device's orientation or position on the body. This is particularly important for devices that are subject to frequent movement, such as fitness trackers and smartwatches. Techniques such as ground plane shaping, dielectric loading, and the use of reflector elements are commonly employed to achieve the desired radiation pattern [18].

Finally, the antenna's efficiency and gain are critical parameters that directly impact the device's communication performance. Efficiency refers to the amount of input power that is radiated as electromagnetic waves, while gain represents the strength of the radiated signal in a specific direction. Achieving high efficiency and gain in wearable antennas is challenging due to the presence of the human body, which can absorb and reflect signals [19]. Engineers use advanced simulation tools and experimental measurements to optimize these parameters and ensure that the antenna performs reliably in real-world conditions.

The frequency range of 2.4–5.8 GHz offers significant advantages for wearable health monitoring devices, providing a balance between compact size, reliable connectivity, and energy efficiency. Microstrip patch antennas are well-suited for this frequency range, offering the versatility and performance required to meet the demands of modern wearable technology.

2.3.3 Designing a Microstrip Patch Antenna for a Fitness Tracker

To better understand the design process, consider the case of a microstrip patch antenna used in a fitness tracker. Fitness trackers are compact, battery-powered wearable devices designed to monitor various health metrics such as physical activity, heart rate, sleep patterns, and other related parameters [8]. These devices leverage Bluetooth technology to wirelessly transmit data to paired smartphones, tablets, or computers, allowing users to track and analyze their health metrics conveniently [20].

The design of a microstrip patch antenna for a fitness tracker involves several critical steps and considerations, which are discussed below in detail.

2.3.3.1 Antenna Size and Shape

Fitness trackers are typically small devices worn on the wrist, meaning the design of the antenna must be compact and lightweight to accommodate the device's size constraints [21]. Microstrip patch antennas are well-suited for such applications due to their low profile, minimal weight, and ability to integrate seamlessly into the device's form factor. The antenna's design can be further optimized to fit within the limited space available without compromising its performance.

The antenna shape, whether rectangular, circular, or customized, is determined by the available space and performance requirements. For example, a rectangular patch is often preferred due to its simplicity in design and ease of fabrication. However, circular or other unconventional shapes may be chosen if they better fit the device's geometry or aesthetic design. Advanced simulation tools such as HFSS or CST Studio are typically used to refine the shape and size of the antenna to ensure optimal performance.

2.3.3.2 Resonant Frequency

The resonant frequency of the antenna is a key parameter in ensuring effective wireless communication. For fitness trackers, the antenna is designed to resonate at the 2.4 GHz ISM (Industrial, Scientific, and Medical) band [22]. This frequency band is commonly used for Bluetooth communication, which is the primary wireless protocol in fitness trackers.

The resonant frequency is determined by the dimensions of the patch antenna and the dielectric properties of the substrate material. The substrate's relative permittivity (ε_r) and thickness play a crucial role in the determination of the patch dimensions. Standard design equations govern the relationship between the resonant frequency and patch dimensions:

where

1. is the effective length of the patch,
2. is the speed of light in a vacuum,
3. is the desired resonant frequency (2.4 GHz),
4. is the effective dielectric constant of the substrate.

$$L = L_{\text{eff}} - 2\,\Delta L \tag{2.1}$$

where ΔL is the length extension due to fringing fields, and is given by:

$$\Delta L = 0.412h\left(\frac{\varepsilon_{\text{eff}} + 0.3}{\varepsilon_{\text{eff}} - 0.258}\right)\left(\frac{\frac{W}{h} + 0.264}{\frac{W}{h} + 0.8}\right) \tag{2.2}$$

By carefully selecting the substrate material and adjusting the dimensions of the patch, designers can achieve precise tuning of the antenna to resonate at the desired frequency. Additionally, parasitic elements or slots may be introduced into the design to fine-tune the resonance or improve bandwidth.

2.3.3.3 Impedance Matching

Impedance matching is a critical aspect of antenna design to ensure maximum power transfer between the antenna and the radio frequency (RF) components of the fitness tracker. Mismatched impedance can lead to signal reflections, reduced efficiency, and poor communication performance [23].

In microstrip patch antennas, impedance matching is typically achieved by adjusting the feedline dimensions or incorporating impedance matching networks. Common techniques include:

1. **Using a Quarter-Wave Transformer:** This method involves designing a feedline with a characteristic impedance that acts as a matching network between the antenna and the transceiver.
2. **Inset Feeding:** By adjusting the inset depth of the feed point into the patch, the input impedance can be fine-tuned to match the transceiver's impedance (usually 50 ohms).
3. **Impedance Matching Networks:** These are external components, such as lumped elements (capacitors and inductors), that compensate for impedance mismatches in the system.

Simulation tools are often employed to analyze and optimize the impedance matching of the antenna design, ensuring minimal signal loss and high performance.

2.3.3.4 Efficiency and Gain

The efficiency of the antenna is critical for ensuring reliable wireless communication without excessive power consumption, which is particularly important for battery-powered devices like fitness trackers. Antenna efficiency is influenced by factors such as:

- **Substrate Material:** Low-loss substrate materials reduce dielectric and conductor losses, improving efficiency.
- **Fabrication Quality:** High-precision fabrication minimizes manufacturing defects that can degrade performance.
- **Design Optimization:** Techniques such as the inclusion of slots or parasitic elements can enhance radiation efficiency.

The gain of the antenna is another important parameter, as it determines the strength of the signal in specific directions. While high gain is desirable for improving communication range, it must be balanced against the size and form factor constraints of the fitness tracker. Directional antennas may be used in some cases, but omnidirectional antennas are generally preferred for wearable devices to ensure consistent performance regardless of orientation.

2.3.3.5 Integration with the Device

Integration of the antenna into the fitness tracker must consider both functional and aesthetic requirements [24]. The antenna is typically positioned on the surface of the device or embedded within its housing. Several factors influence the integration process:

1. **Placement:** The antenna should be placed in a location that minimizes interference from other electronic components, such as the battery, processor, or sensors. For instance, it is often located on the outer surface or edges of the device to maximize signal radiation.
2. **Comfort:** As fitness trackers are worn on the wrist, the antenna design should not compromise user comfort. This includes ensuring that the antenna does not cause irritation or interfere with the wristband.
3. **Material Compatibility:** The materials used in the tracker's housing should not attenuate the antenna's signal. Plastics with low permittivity are often used to encase the antenna to minimize signal loss.
4. **Thermal Management:** The placement and design of the antenna should account for potential heating effects caused by the RF components during operation, ensuring that the device remains safe and comfortable to use.

Advanced techniques such as electromagnetic simulations and prototyping are employed to evaluate the antenna's performance in the actual device environment. This ensures that the final design meets all technical, ergonomic, and aesthetic requirements.

2.3.3.6 Challenges and Solutions

Designing a microstrip patch antenna for a fitness tracker involves addressing several challenges, including:

1. **Miniaturization:** The small size of fitness trackers limits the available space for the antenna. To address this, techniques such as the use of high-permittivity substrates or the incorporation of meandered or folded patch designs can be employed to reduce the antenna's size without compromising performance.
2. **Interference:** The proximity of the antenna to other components can cause electromagnetic interference (EMI), degrading performance. Shielding techniques and careful layout design can mitigate this issue.

3. **Bandwidth:** Microstrip patch antennas typically have narrow bandwidths, which can limit their performance in dynamic environments. Techniques such as the use of slot antennas or multi-resonant structures can enhance bandwidth.
4. **User Interaction:** The presence of the user's body can detune the antenna and affect its performance. This can be addressed by using simulation tools to model the antenna's interaction with the human body and incorporating design features that minimize detuning effects.

This highlights the key design considerations for microstrip patch antennas in wearable health monitoring systems. By addressing challenges such as size constraints, power consumption, and wireless communication requirements, microstrip patch antennas can be optimized to meet the unique demands of fitness trackers. From selecting the right substrate and tuning the resonant frequency to ensuring efficient integration with the device, the design process involves a combination of theoretical calculations, simulations, and practical prototyping.

The successful implementation of microstrip patch antennas in fitness trackers demonstrates their versatility and effectiveness in modern wearable technology, paving the way for further advancements in wireless communication and health monitoring systems.

2.4 Antenna Miniaturization Techniques for Wearable Devices

As wearable devices continue to evolve, there is an increasing demand for smaller, lighter, and more compact antennas that can be seamlessly integrated into the form factor of the device [23]. Antenna miniaturization techniques are crucial for meeting these demands without compromising the performance and efficiency of the antenna. Miniaturization methods often involve reducing the size of the antenna while maintaining the necessary performance characteristics, such as bandwidth, efficiency, and radiation pattern.

The importance of antenna miniaturization lies in its ability to support the development of wearable devices that are not only functional but also comfortable and aesthetically pleasing. Wearable devices like fitness trackers, smartwatches, medical monitors, and augmented reality glasses all rely on efficient antenna designs to enable seamless wireless communication [25]. The challenges of antenna miniaturization arise from the trade-offs between size, performance, and integration with the device. Below, we explore some of the most common techniques used for antenna miniaturization in wearable devices.

2.4.1 Slot Antennas

Slot antennas are a popular choice for miniaturization due to their ability to reduce the effective electrical length of the antenna without significantly altering its physical dimensions. By introducing one or more slots into the antenna structure, the current path is lengthened, allowing the antenna to resonate at a lower frequency. This technique enables designers to create smaller antennas that still meet the desired frequency requirements.

Slot antennas are particularly advantageous for wearable devices because they can be easily integrated into flexible and conformal designs [26]. For example, the slots can be etched onto flexible substrates that conform to the curvature of a smartwatch or fitness tracker. Additionally, slot antennas can be designed to operate in dual-band or multiband configurations, supporting multiple wireless standards such as Bluetooth, Wi-Fi, and LTE. This versatility makes slot antennas a valuable tool for miniaturization.

2.4.2 Fractal Geometries

Fractal geometries involve using complex, self-similar shapes to achieve compact antenna designs. Examples of fractal shapes include the Koch curve, the Sierpinski triangle, and the Minkowski loop [8]. These shapes allow the antenna to occupy less physical space while maintaining a long electrical path, which is essential for achieving the desired resonant frequency.

The use of fractal geometries offers several advantages for wearable devices:

- **Space Efficiency:** Fractal antennas utilize space more efficiently than traditional designs, making them ideal for small wearable devices.
- **Wideband and Multiband Operation:** Fractal shapes can support multiple resonant frequencies, enabling wearable devices to communicate over various wireless protocols.
- **Aesthetic Integration:** The intricate patterns of fractal antennas can be incorporated into the design of wearable devices, enhancing their visual appeal.

Simulation tools are often used to optimize the design of fractal antennas, ensuring that they meet the required performance criteria while fitting within the constraints of the device. Advances in 3D printing and fabrication techniques have further facilitated the development of fractal antennas for wearable applications.

2.4.3 Use of High Dielectric Materials

Materials with high dielectric constants (ε_r) play a critical role in antenna miniaturization. By increasing the effective permittivity of the substrate, the wavelength of the electromagnetic waves within the material is reduced [27]. This allows the antenna to be smaller while still resonating at the desired frequency.

High dielectric materials are particularly useful in applications where space is extremely limited. For example, ceramic-based substrates with high dielectric constants are commonly used in miniaturized antennas for medical implants and other compact wearable devices. However, the use of high dielectric materials also introduces challenges, such as increased dielectric losses and reduced radiation efficiency. To mitigate these issues, designers must carefully select materials that balance permittivity, loss tangent, and fabrication feasibility.

2.4.4 Antenna Folding

Antenna folding involves bending or folding the antenna structure to reduce its overall size. This technique is commonly used in monopole and dipole antennas, where the physical length of the antenna is folded into a more compact configuration. Folded antennas maintain their radiation efficiency and resonant frequency by preserving the electrical length of the current path [28].

One example of antenna folding is the use of meander line designs, where the antenna is folded into a zigzag pattern. Meander line antennas are widely used in wearable devices because they offer:

- **Compact Form Factor:** The folded design minimizes the antenna's footprint, allowing it to fit into tight spaces.
- **Flexibility:** Meander line antennas can be fabricated on flexible substrates, making them suitable for conformal applications.
- **Customizable Designs:** The folding pattern can be adjusted to achieve specific performance characteristics, such as impedance matching and bandwidth enhancement.

In wearable devices, folded antennas are often combined with other miniaturization techniques, such as the use of high dielectric materials or slot designs, to achieve optimal performance.

2.4.5 Planar and Compact Designs

Planar antennas, such as microstrip patch antennas, are inherently compact and well-suited for miniaturization. By optimizing the patch dimensions and substrate properties, planar antennas can be designed to fit within the limited space available in wearable devices [29]. Additional techniques, such as the use of shorting pins or parasitic elements, can further enhance the miniaturization of planar antennas.

One notable advantage of planar antennas is their compatibility with printed circuit board (PCB) fabrication processes. This allows the antenna to be seamlessly integrated into the device's PCB, reducing manufacturing complexity and cost. Planar antennas are also highly customizable, with the ability to support various frequency bands and polarization schemes.

2.4.6 Challenges in Antenna Miniaturization

While antenna miniaturization offers significant benefits for wearable devices, it also presents several challenges that must be addressed to ensure optimal performance:

1. **Bandwidth Reduction:** Miniaturized antennas often have narrower bandwidths, which can limit their ability to operate in dynamic environments. To address this, designers may incorporate techniques such as the use of capacitive loading, slotting, or multi-resonant structures.
2. **Efficiency Losses:** Reducing the size of the antenna can lead to higher losses due to increased resistive and dielectric losses. Advanced materials and fabrication techniques can help mitigate these losses.
3. **User Interaction:** Wearable devices are in close proximity to the user's body, which can detune the antenna and degrade its performance. This challenge can be addressed by modeling the interaction between the antenna and the human body during the design process.
4. **Interference:** The integration of multiple antennas within a small device can lead to electromagnetic interference (EMI). Techniques such as shielding, isolation, and careful layout design can minimize interference.

2.4.7 Emerging Trends in Antenna Miniaturization

The field of antenna miniaturization is constantly evolving, with new technologies and approaches being developed to meet the demands of wearable devices. Some emerging trends include:

1. **Metamaterials:** The use of engineered materials with unique electromagnetic properties, known as metamaterials, enables the design of compact antennas with enhanced performance. Metamaterials can be used to create antennas with negative refractive indices, superlenses, and other advanced characteristics.
2. **3D Printing:** Additive manufacturing techniques, such as 3D printing, offer new possibilities for creating complex antenna structures with high precision. This is particularly useful for fabricating conformal and customized antennas for wearable devices.
3. **Flexible and Stretchable Antennas:** Advances in materials science have enabled the development of flexible and stretchable antennas that can conform to the shape of the human body. These antennas are ideal for applications such as smart textiles, medical monitoring, and fitness tracking.
4. **AI-Driven Design:** Artificial intelligence (AI) and machine learning (ML) are being used to optimize antenna designs. By analyzing vast datasets and exploring complex design spaces, AI can identify innovative solutions for miniaturization and performance enhancement.

Antenna miniaturization is a critical aspect of wearable device design, enabling the integration of efficient antennas within the strict size and weight constraints of these devices. Techniques such as slot antennas, fractal geometries, high dielectric materials, antenna folding, and planar designs provide designers with a wide range of options for achieving compact and high-performance antenna solutions.

By addressing the challenges of miniaturization and leveraging emerging technologies, researchers and engineers can continue to push the boundaries of wearable device capabilities. The future of antenna miniaturization holds great promise, with innovations such as metamaterials, 3D printing, and AI-driven design paving the way for even smaller, more efficient, and versatile antennas. These advancements will play a key role in shaping the next generation of wearable technology, enhancing their functionality, comfort, and overall user experience.

2.5 Enhancing Antenna Efficiency and Radiation Pattern for Body-Worn Applications

The efficiency and radiation pattern of an antenna are key factors in determining the quality of wireless communication in body-worn applications [30]. These factors can be affected by the proximity of the human body, which influences the antenna's performance. Understanding how to enhance antenna efficiency and optimize radiation patterns is essential for ensuring reliable and effective communication.

2.5.1 Methods to Improve Antenna Efficiency and Radiation Pattern

To improve the efficiency and radiation pattern of microstrip patch antennas for wearable devices, several techniques can be employed:

1. **Antenna Structure Optimization:** The shape and size of the patch can be adjusted to achieve better impedance matching and minimize signal losses. The use of different patch geometries, such as circular, square, or triangular shapes, can help optimize the radiation pattern.
2. **Ground Plane Modifications:** The ground plane size and shape can significantly affect the antenna's radiation pattern. By optimizing the ground plane, one can reduce losses and improve the radiation efficiency. For example, a larger ground plane can help reduce the effects of body proximity.
3. **Use of Metamaterials:** Metamaterials are engineered materials with properties not found in naturally occurring substances. By incorporating metamaterials into the antenna design, it is possible to improve radiation efficiency and reduce size, leading to better performance in wearable applications.
4. **Antenna Placement on the Body:** The position of the antenna on the body can also impact its radiation pattern and efficiency. Antennas placed near the skin or on areas with high electrical conductivity, such as the chest or wrist, can experience detuning effects. Designing antennas that minimize the influence of the human body is important for achieving better radiation patterns and efficiency.
5. **Antenna Loading:** Loading techniques, such as incorporating inductive or capacitive elements, can improve the efficiency and bandwidth of the antenna. This method helps in enhancing the antenna's performance without increasing its size.

By employing these methods, wearable devices can achieve improved communication performance, ensuring better signal transmission and reception in body-worn applications.

2.5.2 Challenges Related to Antenna Detuning

Antenna detuning refers to the shift in the resonant frequency of an antenna due to environmental factors, such as the presence of the human body. In wearable health monitoring devices, the body's dielectric properties can cause detuning, leading to reduced antenna efficiency and performance [31].

Several factors contribute to antenna detuning in wearable devices:

1. **Body Proximity:** The body's conductive nature can change the effective impedance and resonant frequency of the antenna. When the antenna is placed close to the body, it may experience detuning, resulting in poor performance.

2. **Body Movement:** The movement of the body, including bending or stretching of the skin, can cause mechanical stress on the antenna, leading to changes in its geometry and detuning. This is particularly relevant for wearables such as fitness trackers and smartwatches, which are subject to continuous motion.
3. **Human Tissue Interaction:** The dielectric properties of human tissues vary across the body, influencing the way electromagnetic waves propagate. The absorption and reflection of radio waves by human tissues can affect the antenna's radiation efficiency and radiation pattern.

To mitigate detuning, designers often use techniques such as body effect compensation, optimization of antenna placement, and adaptive tuning circuits to ensure that the antenna operates effectively in the presence of the human body.

2.5.3 *Wireless Communication in the Human Body*

Wireless communication within the human body, also referred to as Body Area Networks (BANs), represents a rapidly growing field with significant potential for healthcare and medical monitoring (see Fig. 2.2). BANs utilize wireless technologies to connect devices on or within the human body, enabling real-time monitoring of various physiological parameters. However, one of the most significant challenges in establishing effective communication within the body is the interference that the body itself poses to radio waves. The human body, with its high water content, behaves as a large, highly conductive medium, which leads to substantial signal attenuation and degradation as radio waves pass through [32].

The human body is predominantly composed of water, with an average of 60–70% of body weight being water, which makes it an excellent conductor of electrical signals. This conductive property results in high absorption and scattering of electromagnetic waves, particularly at frequencies that are typically used in

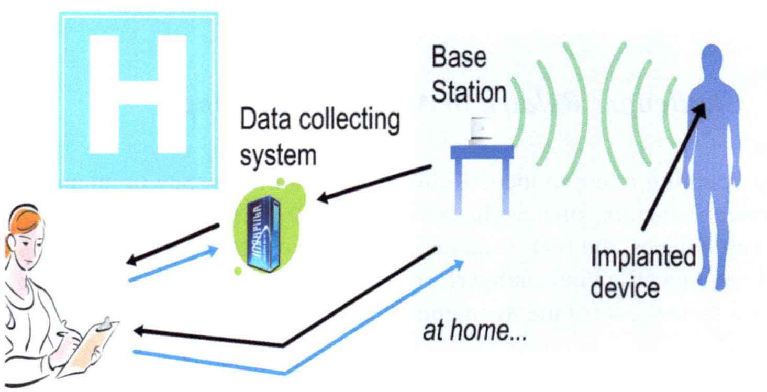

Fig. 2.2 A typical home healthcare system featuring a wireless implantable device operating within a Wireless Body Area Network (WBAN) [32]

communication systems. As a result, conventional wireless communication techniques may not perform optimally when the signal must propagate through the human body, leading to severe signal loss.

Additionally, the presence of other materials such as bones, muscles, and fat further complicates signal transmission. The variation in tissue types introduces different levels of attenuation and reflection, making it difficult to predict how a signal will behave in a given scenario. For instance, bone, being denser and more electrically conductive, can cause more signal loss than fat tissue. Similarly, skin and muscle layers can also absorb and scatter electromagnetic waves differently, depending on the frequency used.

Given these challenges, several strategies have been developed to mitigate the effects of the human body on wireless communication. Some of the most widely adopted techniques include:

1. **Short-range communication**: Short-range communication technologies, such as Bluetooth Low Energy (BLE) and Zigbee, are commonly used in wearable devices to overcome the limitations imposed by body interference. These technologies are designed to operate at lower power levels, with communication ranges typically extending from a few centimeters to a few meters. By focusing on short-range communication, the potential for signal attenuation caused by the body can be minimized, as the signal does not need to propagate over long distances [33]. BLE, for example, is widely used in wearable health monitoring devices because of its low energy consumption, which is ideal for devices that require long battery life while maintaining reliable connectivity.

   The use of short-range communication standards also benefits from their ability to use unlicensed frequency bands, such as the 2.4 GHz ISM band, which is widely available and less prone to regulatory constraints. While these frequencies can still experience attenuation in the presence of the human body, they provide a good balance between range and signal quality for wearable applications.
2. **Near-field communication**: Near-field communication (NFC) is another promising technique for in-body communication systems, particularly in medical implants and health monitoring devices. NFC technology allows for communication over very short distances (typically less than 10 cm) and operates at low power levels, making it suitable for battery-powered devices that are implanted within the body or worn on the skin. NFC relies on electromagnetic induction to establish communication between the device and an external reader, without the need for a direct line of sight or complex signal processing.

   The main advantage of NFC is its ability to provide secure, short-range communication without the need for high-power transmissions. This is particularly beneficial in medical applications, where low-power consumption and biocompatibility are essential. Additionally, NFC can be used for tasks such as data transfer, device synchronization, and power harvesting in health monitoring systems.

3. **Optimized frequencies**: One way to address the challenge of signal attenuation in the human body is to operate at higher frequencies, such as those in the 2.4–5.8 GHz ISM bands. These higher frequencies offer better propagation characteristics in short-range communication scenarios, as the signal is less affected by the body's interference when traveling over shorter distances. The 2.4 GHz frequency, in particular, is widely used in wireless communication systems, including Wi-Fi, Bluetooth, and Zigbee, due to its favorable balance between range, data rate, and signal penetration.

   Operating at higher frequencies also allows for greater bandwidth, enabling faster data transfer rates, which is essential for applications such as continuous health monitoring and real-time data transmission. However, it is necessary to note that higher frequencies are still subject to some level of attenuation when passing through the human body, and their ability to penetrate specific tissues may be limited. Thus, careful optimization of frequency selection is necessary to ensure reliable communication in wearable health devices.
4. **Body-Centric Wireless Networks**: Beyond individual devices, there is growing interest in creating body-centric wireless networks that connect multiple wearable devices on the human body, forming a seamless network for health monitoring and other applications. These networks can include sensors, actuators, and communication devices that are integrated into clothing, patches, or medical implants. A key challenge in these networks is ensuring that the communication between devices remains stable as the body moves, changes position, and interacts with the environment. Special techniques such as adaptive beamforming, frequency hopping, and multi-path propagation modeling are being developed to address these challenges and improve the reliability of body-centric communication systems [34].

2.5.4 Consistent Performance in Wearable Devices

Achieving consistent performance in wearable devices requires overcoming various environmental challenges, particularly those caused by body movements, physical exertion, and fluctuations in the surrounding environment. Wearable health devices often rely on wireless communication to transmit data, monitor vital signs, or trigger alerts, and maintaining consistent and reliable performance is crucial for ensuring the effectiveness of these devices in real-world scenarios.

Wearable devices, such as smartwatches, fitness trackers, and medical sensors, are subject to dynamic movements as they are worn on the body. The position and orientation of the device can change significantly during activities such as walking, running, or even sleeping. These movements can alter the performance of the device's antenna and communication system, leading to fluctuations in signal strength and data transmission reliability [35].

To address these challenges, wearable device designers focus on optimizing the antenna design for robustness. A well-designed antenna should be able to perform reliably under a wide range of conditions, including during body movement, temperature changes, and interference from other devices. Several strategies have been employed to improve the performance of antennas in wearable devices:

1. **Adaptive tuning methods**: One effective technique for ensuring consistent performance is the use of adaptive tuning methods, which allow the antenna to adjust its performance based on environmental factors. Adaptive tuning involves dynamically adjusting the antenna's impedance, frequency, or other parameters to optimize signal reception and transmission. For instance, an adaptive antenna could compensate for changes in the body's position or movement, ensuring that the device maintains a strong and stable connection with the wireless network or other devices.

   Adaptive tuning can also be used to mitigate interference from other electronic devices, such as smartphones or wireless routers, which may introduce noise or disrupt communication. By adjusting the antenna's frequency or power settings, the wearable device can minimize interference and ensure reliable communication in diverse environments.
2. **Wearable antenna design considerations**: The physical design of the antenna is also crucial for maintaining consistent performance. Antennas for wearable devices must be compact, flexible, and lightweight while still offering sufficient signal strength and bandwidth. Designers often use conformal antenna designs that allow the antenna to bend and conform to the contours of the body without sacrificing performance. These flexible antennas can be integrated into clothing, wristbands, or other wearable accessories, ensuring that the device remains functional and comfortable during use.

   Moreover, the material selection for the antenna is critical to its performance. Conductive materials such as silver, copper, or gold are commonly used to create efficient antennas, while flexible substrates such as polydimethylsiloxane (PDMS) or polyethylene terephthalate (PET) offer the necessary flexibility and durability for wearable applications. These materials allow the antenna to maintain consistent performance even under conditions of physical movement and stress.
3. **Energy harvesting**: Another strategy for improving the performance of wearable devices is the incorporation of energy harvesting technologies. Many wearable devices rely on batteries for power, and ensuring that the device can operate consistently over extended periods is a significant challenge. Energy harvesting techniques, such as piezoelectric or thermoelectric systems, can convert ambient energy (such as body movement or heat) into electrical power, extending the device's battery life and ensuring consistent performance without frequent recharging.

   Energy harvesting can be particularly useful in health monitoring applications, where continuous data collection is necessary. By integrating energy harvesting systems into the device, designers can reduce the need for frequent charging and ensure that the device remains operational even during long-term wear.

2.6 Materials for Biocompatible and Flexible Microstrip Patch Antennas

The development of wearable health monitoring devices, especially those that incorporate wireless communication, requires careful consideration of the materials used in their construction. In particular, the materials used for microstrip patch antennas in wearable devices must meet two crucial requirements: biocompatibility and flexibility [36].

1. **Biocompatibility**: Biocompatibility refers to the ability of a material to interact with the human body without causing adverse effects. For wearable devices, this is a critical consideration, as the materials used in the device must be safe for prolonged contact with the skin. Materials that are not biocompatible can cause irritation, allergic reactions, or other harmful effects when in contact with the skin or other body tissues. Therefore, biocompatible materials must be chosen to ensure that the device can be worn safely for extended periods, especially in medical applications where the device may be in constant contact with the body [37].

   Conductive materials used in wearable antennas, such as gold, silver, or copper, are generally biocompatible, as they do not react with the skin or body fluids. However, other materials, such as certain polymers or adhesives, may require further testing to ensure that they do not cause irritation or other health issues.
2. **Flexibility**: Flexibility is another essential characteristic for materials used in wearable devices, as the device must be able to bend, stretch, and conform to the body's movements. Flexible antennas are particularly important for ensuring consistent performance in wearable devices, as they can maintain signal strength and quality even when subjected to physical movement and deformation [38].

   Flexible materials such as polydimethylsiloxane (PDMS) and polyethylene terephthalate (PET) are commonly used as substrates for wearable antennas. PDMS, in particular, is a silicone-based material known for its high flexibility, biocompatibility, and durability. PET, a polyester-based material, is also flexible and lightweight, making it ideal for wearable applications. These materials provide the necessary flexibility for integration into clothing, wristbands, and other wearable accessories, allowing the antenna to maintain performance while being worn on the body.

   Conductive materials used in flexible antennas, such as silver ink or copper traces, must also be flexible and durable to ensure that the antenna maintains its performance over time. Silver ink, for example, can be printed onto flexible substrates using techniques such as inkjet printing, allowing for the creation of lightweight, flexible antennas that can be easily integrated into wearable devices.
3. **Wearable antenna performance**: The choice of materials used for wearable antennas has a direct impact on the performance of the device. Materials must be selected not only for their biocompatibility and flexibility but also for their ability to support the necessary electromagnetic properties for efficient wireless

communication. High-quality conductive materials ensure that the antenna can transmit and receive signals effectively, while flexible substrates allow the antenna to maintain its shape and function even when the device is in motion [39].

2.7 Coexistence and Interference Mitigation in Wearable Antenna Design

The integration of wearable health devices in everyday life has ushered in a new era of continuous health monitoring. These devices, ranging from fitness trackers to more sophisticated biomedical wearables, rely on wireless communication to transmit data to smartphones, cloud servers, or medical facilities. However, one of the most significant challenges these devices face is operating in crowded radio frequency (RF) environments. In these environments, numerous wireless devices, such as smartphones, Wi-Fi routers, Bluetooth-enabled devices, and medical equipment, transmit and receive data over the same or overlapping frequency bands. This can lead to interference, which affects the performance of wearable antennas and the overall functionality of these devices.

The importance of mitigating interference in wearable antenna design cannot be overstated, as poor performance or failure in transmission can lead to inaccurate readings or loss of critical data. As wearable health devices become more commonplace, the ability to design antennas that can effectively coexist with other devices, without degrading performance, has become a key area of research in antenna design. This section will explore the challenges of coexistence with other wireless devices, the various interference mitigation techniques, and methods for ensuring optimal performance in crowded RF environments [40].

2.7.1 Challenges Related to Coexistence with Other Wireless Devices

In modern environments, many wireless devices operate within the same frequency bands. This can lead to interference, where signals from one device overlap with signals from another device, causing degradation in performance. The presence of multiple wireless devices in close proximity, especially in homes, offices, and hospitals, further exacerbates this issue. Interference can manifest in several forms, including signal attenuation, phase distortion, and even complete signal loss, which can compromise the functionality of wearable health devices.

One of the most significant challenges in wearable antenna design is the limited frequency spectrum available for wireless communication. Frequencies in the radio spectrum are regulated and allocated for specific uses, such as mobile communications, Wi-Fi, Bluetooth, and medical devices. However, the demand for wireless communication continues to increase as more devices are connected to the Internet of Things (IoT), further congesting the available spectrum. As a result, wearable devices must be designed to share these limited frequency bands efficiently with other devices without causing interference or performance degradation [41].

Another challenge comes from the proximity of wearable devices to other wireless equipment. Wearable health devices are typically worn on the body or attached to clothing, meaning they are often in close proximity to other wireless devices, such as smartphones or wireless routers. This proximity can lead to signal coupling, where unwanted signals from nearby devices affect the wearable antenna's performance. Additionally, the small size and form factor of wearable devices limit the space available for antenna design, making it even more challenging to implement effective coexistence strategies without compromising the device's size, weight, or comfort.

The growing trend of incorporating multiple wireless technologies into a single device, such as a wearable health monitor that uses Bluetooth for short-range communication and Wi-Fi for long-range connectivity, also presents challenges. When multiple wireless interfaces are used simultaneously, it becomes even more difficult to ensure that they do not interfere with one another. For instance, when Bluetooth and Wi-Fi operate in overlapping frequency bands, interference can occur, leading to reduced data throughput or complete communication failure.

2.7.2 Interference Mitigation in Crowded RF Environments

Interference mitigation is crucial for ensuring reliable communication in crowded RF environments. Several techniques have been developed to address interference and improve the performance of wearable antennas, especially in environments where multiple devices are transmitting simultaneously.

Frequency Hopping: One of the most commonly used techniques to mitigate interference in crowded environments is frequency hopping. This technique involves rapidly changing the frequency at which a device transmits its signal, according to a predefined sequence. By hopping between frequencies, the device can avoid periods of interference caused by other devices operating on the same frequency. Bluetooth, for example, employs frequency hopping to avoid interference with other devices in the 2.4 GHz ISM (Industrial, Scientific, and Medical) band. Frequency hopping spreads the signal across multiple frequencies, making it less susceptible to interference from any device [42].

Spectrum Sharing: Spectrum sharing is another important technique for managing interference. This approach allows multiple devices to share the same frequency band by coordinating their use of the spectrum. Spectrum sharing can be implemented using various strategies, such as time-division multiplexing (TDM), where devices take turns transmitting on the same frequency, or space-division multiplexing (SDM), where devices use different spatial channels to communicate. In the case of wearable devices, spectrum sharing can help prevent signal interference, especially when several devices are operating in close proximity. Dynamic spectrum access (DSA) is an emerging approach that allows devices to intelligently select available spectrum bands based on current usage, further optimizing spectrum utilization.

Dynamic Frequency Selection (DFS): Dynamic frequency selection (DFS) is a method that enables devices to automatically switch to a different frequency band when interference is detected on their current band. DFS is commonly used in Wi-Fi networks to avoid interference with radar systems operating in the same frequency band. Wearable devices equipped with DFS capabilities can detect and prevent interference from other devices by adjusting their transmission frequency in real time. This dynamic approach enhances the performance of wearable antennas in environments with dense RF traffic [43].

Cooperative Communication: In some cases, multiple wearable devices can work together to mitigate interference through cooperative communication. By exchanging information about the communication environment, devices can adjust their transmission power or frequency to minimize interference. This cooperative approach can be particularly useful in healthcare settings, where wearable health devices must coexist with medical equipment and other wireless devices. For instance, in a hospital, a group of wearable health monitors might cooperate to avoid interference with critical medical equipment, such as infusion pumps or patient monitoring systems.

2.7.3 Techniques for Filtering and Isolation

To further reduce interference, wearable antennas often incorporate filters and isolators into their design. These components help isolate the wearable device from unwanted signals and ensure that only the desired frequencies are transmitted or received.

Filters: Filters are electronic components used to block unwanted frequencies while allowing the desired frequencies to pass through. In the context of wearable antenna design, filters are used to reduce interference from signals outside the operating frequency band of the device. For example, a low-pass filter can block high-frequency interference, while a band-pass filter can ensure that only signals within a specific frequency range are allowed to pass through. By incorporating filters into the antenna design, engineers can enhance the signal quality and reduce the likelihood of interference from nearby devices operating on different frequencies.

Isolators: Isolators are passive components that prevent signals from interfering with each other by isolating different parts of the system. In wearable antenna design, isolators can be used in order to avoid mutual coupling between the antenna and other components, such as the body or surrounding electronics. This ensures that the wearable device operates without interference from nearby objects or signals. Isolators can be particularly useful when designing antennas for biomedical wearables, where the antenna may be in close proximity to the human body, which can introduce unwanted signal reflections or absorption.

Electromagnetic Shielding: In addition to filters and isolators, electromagnetic shielding is another technique used to reduce interference. Shielding involves enclosing sensitive electronic components, such as antennas and circuits, in materials that block or absorb electromagnetic radiation. Shielding can help protect the wearable antenna from external interference, such as signals from nearby wireless devices or environmental noise. Common materials used for electromagnetic shielding include conductive metals, such as copper or aluminum, which effectively block unwanted RF signals [44].

Impedance Matching: Impedance matching is a critical aspect of antenna design that ensures maximum power transfer between the antenna and the device's radio circuit. Mismatched impedance can lead to signal reflections and loss of power, which can degrade the performance of the wearable device. By carefully designing the antenna's impedance characteristics to match those of the radio circuit, engineers can minimize signal loss and improve overall performance. Impedance matching is critical in environments with high interference, as it helps to maintain signal integrity even in the presence of noise or unwanted signals.

2.8 Antenna Integration in Smart Textiles and Biomedical Wearables

The integration of antennas into smart textiles and biomedical wearables is an exciting development in healthcare and wearable technology. These devices use textile-based materials, such as fabrics and fibers, to embed sensors and antennas directly into clothing or accessories, enabling continuous monitoring of vital signs such as heart rate, body temperature, and movement. Smart textiles and biomedical wearables offer several advantages over traditional wearable devices, including increased comfort, flexibility, and unobtrusiveness.

However, integrating antennas into textiles presents several unique challenges that must be carefully addressed to ensure both performance and comfort. The design of antennas for smart textiles requires a deep understanding of the materials used in the textiles, as well as the mechanical properties and form factors of the wearable devices. Below, we discuss the key considerations for antenna integration in smart textiles and biomedical wearables.

Materials for Smart Textiles: One of the primary challenges in antenna integration is selecting the appropriate materials for both the antenna and the textile substrate. Conductive materials such as copper, silver, and conductive polymers are commonly used to create antennas that can be woven or printed into fabrics. These materials must have the necessary electrical conductivity to support RF communication while maintaining flexibility and durability. Additionally, the materials should be lightweight and comfortable to wear, as they will be in constant contact with the skin [45].

In some cases, antennas can be embedded into the fabric using printing techniques such as inkjet printing or screen printing, which allow for the creation of flexible, low-cost antennas that can be seamlessly integrated into clothing. The choice of materials and manufacturing techniques will depend on the specific application of the wearable device, as well as the desired balance between performance and comfort.

Form Factors and Comfort: Since wearable devices are typically worn for extended periods, comfort is a critical factor in their design. Wearable antennas must be designed to fit comfortably into the fabric without causing discomfort or restricting movement. The antenna's form factor must be carefully considered to ensure that it does not interfere with the wearer's daily activities. For example, in a medical wearable device designed to monitor heart rate, the antenna should be small, flexible, and lightweight, allowing the device to be worn under clothing without causing discomfort.

Smart textiles also need to be breathable and moisture-wicking to ensure comfort over long periods of wear. The integration of antennas must not compromise these properties, as wearable health devices need to maintain a close fit to the skin for accurate sensing while also ensuring that the wearer does not feel restricted or uncomfortable.

Performance Considerations: While comfort is important, the performance of the wearable antenna cannot be sacrificed. Antennas embedded in smart textiles must deliver reliable RF communication, with minimal signal loss and interference. Factors such as antenna size, placement, and orientation relative to the body must be carefully optimized to achieve optimal performance. Additionally, the impact of body movement on the antenna's performance should be considered, as wearable devices are subject to constant motion during daily activities [46].

Biomedical Wearables and Health Monitoring: In the case of biomedical wearables, antennas must be designed to operate in the presence of complex human body environments. The body itself can introduce additional challenges, such as signal absorption, reflection, and interference from bodily fluids or tissues. Biomedical wearables, such as continuous glucose monitors or ECG sensors, require precise and stable signal transmission to ensure the accuracy of health data. Antennas designed for these devices must be optimized to minimize losses due to the human body and to ensure reliable communication with external devices or monitoring systems.

The integration of smart textiles and biomedical wearables represents a growing field with immense potential to revolutionize healthcare and personal wellness. By combining the advantages of flexible, wearable antennas with the convenience and comfort of textiles, these devices offer a unique opportunity to enhance the continuous monitoring of health metrics, leading to better-informed medical decisions and improved quality of life for users.

2.9 Open Research Challenges in Wearable Antenna Design

Wearable antennas play a key role in health and communication devices, operating in dynamic and congested environments.

2.9.1 Improving Antenna Miniaturization without Compromising Performance

One of the fundamental challenges in wearable antenna design is miniaturization. Wearable devices must be compact and lightweight to ensure comfort and practicality. However, as the antenna size decreases, maintaining or improving the performance of the antenna becomes increasingly difficult. The performance of an antenna is primarily determined by its physical dimensions, and smaller antennas typically suffer from reduced efficiency, bandwidth, and radiation characteristics [47].

The process of miniaturization often leads to challenges such as increased antenna resistance and reduced quality factor (Q-factor), which can cause signal attenuation and loss of power. Furthermore, the proximity of the antenna to the human body can affect its performance due to the body's dielectric properties. The human body is an effective absorber of electromagnetic energy, and when a wearable antenna is worn close to the skin, it can alter the radiation pattern and efficiency of the antenna, particularly at specific frequencies.

The research community is exploring several strategies to overcome these limitations. One approach involves using metamaterials and metasurfaces to create miniaturized antennas that retain high performance. Metamaterials are engineered materials with properties not typically found in nature, and they can be designed to have specific electromagnetic characteristics that enable the creation of small antennas with enhanced radiation efficiency and bandwidth. Another approach is the development of multiband or broadband antennas that operate efficiently across a wide range of frequencies, allowing wearable devices to support multiple communication protocols without compromising performance.

2.9.2 Development of Biocompatible Materials for Wearable Antennas

Biocompatibility is another critical challenge in wearable antenna design. Wearable devices, especially those intended for long-term use, must be made from materials that are not only effective at supporting RF communication but also safe and comfortable for prolonged contact with the human body. The use of biocompatible materials ensures that the wearable device does not cause skin irritation, allergic reactions, or other health issues.

Traditional antenna materials, such as metals and conductive polymers, may not be suitable for direct skin contact, as they can cause discomfort or react with sweat and oils on the skin. Additionally, many metals can corrode over time, reducing the antenna's performance. To address these issues, researchers are exploring new biocompatible materials that can meet the unique demands of wearable antennas [48].

Materials such as flexible conductive polymers, carbon-based materials like graphene, and metallic coatings with high corrosion resistance are being explored for wearable antenna, Materials like graphene and flexible polymers offer conductivity, flexibility, and biocompatibility, making them ideal for wearable antennas.

In addition to the physical properties of the materials, researchers are also concerned with their environmental impact. The use of sustainable materials and manufacturing processes is becoming increasingly important, especially as wearable devices are often disposed of after a relatively short lifespan. Developing biocompatible materials that are also environmentally friendly is a key area of research in the field of wearable antennas.

2.9.3 Optimizing Antenna Designs for Body-Centric Communication

Wearable devices, particularly health monitoring devices, require antennas that can function effectively in body-centric environments. Body-centric communication (BCC) refers to communication systems that involve the human body as a medium for transmitting electromagnetic signals. In wearable devices, the body's presence can significantly affect antenna performance, due to factors such as body shape, tissue density, and proximity to the antenna [49].

When a wearable antenna is placed near the body, it can experience detuning, which occurs when the body absorbs or scatters the electromagnetic waves, leading to a reduction in the antenna's radiation efficiency. The body acts as a lossy medium, especially at frequencies used in wearable devices, such as those in the ISM bands (e.g., 2.4 GHz for Bluetooth and Wi-Fi). The result is that the antenna may need to be larger to compensate for the body's effect on its performance, which contradicts the need for miniaturization.

One challenge in designing antennas for body-centric communication is the development of antennas that can adapt to the varying shape and size of the human body.

Another approach is to design antennas that are optimized for specific body parts or areas of the body. For example, wearable health devices such as smartwatches or fitness trackers require antennas that are optimized for wrist communication, while biomedical sensors that monitor heart rate or glucose levels may require antennas that are optimized for the chest area. By tailoring the antenna design to specific body regions, it is possible to improve the performance and efficiency of body-centric communication systems [50].

Additionally, researchers are working on designing wearable antennas that minimize their interaction with the body. This includes using techniques such as the placement of the antenna away from the skin or the use of materials that reduce the impact of body tissue on the antenna's performance. Techniques like the use of dielectric coatings or the incorporation of air gaps between the antenna and the body can help reduce the absorption of electromagnetic waves and improve overall antenna performance.

2.9.4 Overcoming Interference Issues in Crowded RF Environments

As wearable devices become more widespread, the challenge of interference in crowded RF environments becomes increasingly significant. Wearable health devices must coexist with a variety of other wireless systems, such as smartphones, Wi-Fi routers, Bluetooth devices, and even medical equipment in healthcare settings. In these environments, interference from other devices operating on the same or adjacent frequencies can degrade the performance of wearable antennas, leading to reduced signal quality, dropped connections, and compromised functionality.

Interference issues are exacerbated by the limited availability of spectrum in popular frequency bands. The ISM bands, which are commonly used for Bluetooth, Wi-Fi, and other wireless communication protocols, are heavily congested, with multiple devices often transmitting simultaneously. Furthermore, the RF spectrum is not uniformly distributed, and there are regulatory restrictions on which frequency bands can be used for specific applications. These limitations make it difficult for wearable antennas to maintain reliable communication in dense RF environments [51].

Several techniques are being researched to mitigate interference in crowded RF environments. One of the most promising solutions is the use of dynamic spectrum access (DSA), which enables wearable devices to select available frequency bands based on current usage adaptively. DSA allows devices to operate in frequency bands that are underutilized, thus avoiding interference with heavily congested bands. Additionally, dynamic frequency selection (DFS) can be used to enable devices to automatically switch to a different frequency when interference is detected.

Another approach to mitigating interference is the use of advanced filtering techniques. Filters can be employed to block unwanted signals or reduce noise from external sources. For example, band-pass filters can ensure that only signals within a specific frequency range are transmitted or received, preventing interference from other devices operating on adjacent frequencies. In addition to filters, the use of isolation techniques, such as spatial diversity and polarization diversity, can help reduce the impact of interference on wearable antenna performance.

Cooperative communication is another potential solution for managing interference. In cooperative communication, multiple devices can share information about the interference environment and work together to avoid interference. This can be particularly useful in scenarios where multiple wearable devices are used in close proximity, such as in a healthcare setting where patients may be wearing multiple health monitoring devices. By coordinating their transmission frequencies and power levels, these devices can reduce the likelihood of interference and improve overall system performance [52–54].

The integration of artificial intelligence (AI) and machine learning (ML) techniques into wearable antenna design can provide adaptive solutions for interference mitigation. Wearable antenna design is an evolving field with numerous challenges and opportunities for research and innovation. Miniaturization, biocompatibility, and interference issues will drive wearable antenna innovation. As technology progresses, the development of novel materials, computational methods, and adaptive techniques will pave the way for next-generation wearable antennas that are smaller, more efficient, and more capable of operating in complex and crowded RF environments.

2.10 Summary

Chapter 2 provides an in-depth exploration of microstrip patch antennas in the context of wearable health monitoring devices. The chapter addresses key considerations for designing antennas that can be seamlessly integrated into wearable systems while maintaining high efficiency and performance. Through a combination of theoretical concepts and practical insights, the chapter outlines various design principles, challenges, and techniques aimed at optimizing antenna functionality in wearable devices.

The chapter begins with an introduction to the role of microstrip patch antennas in wireless communication for wearable health monitoring. Microstrip patch antennas are particularly suitable for this purpose due to their compact size, lightweight nature, and ease of integration into wearable devices. These antennas must be designed to perform well within the constraints of wearable technology, where space, power, and efficiency are of paramount importance.

A significant focus of the chapter is on the various wireless communication systems that are commonly employed in wearable health devices. Wireless communication, particularly using technologies like Bluetooth Low Energy (BLE), Zigbee, and Wi-Fi, is integral to ensuring that wearable devices can transmit health data reliably and efficiently. The chapter investigates into the frequency bands commonly used for wearable communications, including the 2.4–5.8 GHz range, and explores the design challenges associated with ensuring a stable and high-performance wireless link despite the presence of the human body.

The design considerations for wearable antennas are crucial to achieving optimal performance. These include factors like the size, shape, and material properties of the antenna, as well as its placement on the body. Microstrip patch antennas are favored for their planar structure, which allows them to be integrated easily into wearable devices such as fitness trackers, smartwatches, and health monitoring patches. The chapter emphasizes the importance of selecting the appropriate substrate material and patch geometry to balance performance, flexibility, and biocompatibility.

The chapter also explores antenna miniaturization techniques, which are critical for fitting antennas into small wearable devices without compromising their performance. Techniques such as the use of slot antennas, fractal geometries, high dielectric materials, and compact planar designs are discussed in detail. These methods help reduce the physical size of the antenna while maintaining sufficient radiation efficiency and bandwidth for effective communication.

One of the key challenges in wearable antenna design is ensuring high antenna efficiency and maintaining an optimal radiation pattern, especially given the presence of the human body. The body's conductive nature can interfere with the antenna's performance, leading to issues like detuning, which can negatively affect the antenna's resonant frequency and radiation characteristics. The chapter covers various methods to improve antenna efficiency, such as antenna structure optimization, ground plane modifications, and the use of metamaterials, as well as strategies for minimizing the impact of body proximity on performance.

Detuning is a major concern in wearable antenna design, as the human body can alter the impedance of the antenna, leading to performance degradation. The chapter highlights techniques for mitigating detuning, such as using adaptive tuning circuits, optimizing antenna placement, and employing body effect compensation strategies. These methods ensure that the antenna remains efficient and performs reliably even in dynamic environments where the body is in motion.

Wireless communication in the human body, or Body Area Networks (BANs), is another important aspect covered in this chapter. The chapter addresses the challenges of signal attenuation caused by the human body's high water content and conductivity. To overcome these issues, wearable devices often operate using short-range communication technologies and optimized frequencies. The use of near-field communication (NFC) and other low-power communication protocols is explored as a means to facilitate efficient communication in body-worn devices.

The chapter also discusses the materials used in the design of biocompatible and flexible microstrip patch antennas. For wearable health devices, the materials must not only be flexible and lightweight but also biocompatible to avoid adverse reactions when in contact with the skin. Flexible substrates like polydimethylsiloxane (PDMS) and polyethylene terephthalate (PET) are highlighted for their suitability in wearable applications, while biocompatible conductive materials such as silver and gold are used for their excellent conductivity and non-reactive properties.

Another key topic covered is the coexistence of wearable antennas with other wireless devices. In crowded RF environments, interference from other devices can degrade the performance of wearable antennas. The chapter explores techniques for interference mitigation, such as frequency hopping, spectrum sharing, and dynamic frequency selection (DFS), which help ensure that wearable devices can coexist with other wireless technologies without significant performance loss.

The integration of antennas into smart textiles and biomedical wearables is also discussed. Smart textiles and biomedical wearables are revolutionizing healthcare by embedding sensors and antennas into fabrics, enabling continuous monitoring of vital signs. The chapter examines the challenges and opportunities of integrating antennas into these flexible and wearable technologies, which require careful consideration of material selection, design, and comfort.

Finally, the chapter concludes with an overview of the open research challenges in wearable antenna design. These challenges include improving antenna miniaturization, developing new biocompatible materials, optimizing antenna designs for dynamic body environments, and overcoming interference in crowded RF bands. Addressing these challenges will be crucial for the continued development of advanced wearable health monitoring systems.

This provides a comprehensive overview of the principles and challenges involved in the design of microstrip patch antennas for wearable health monitoring applications. By focusing on antenna miniaturization, efficiency, radiation pattern optimization, material selection, and interference mitigation, the chapter outlines the key strategies for developing wearable devices that can effectively monitor health parameters while ensuring comfort, reliability, and performance.

References

1. C.V. Anikwe, H.F. Nweke, A.C. Ikegwu, C.A. Egwuonwu, F.U. Onu, U.R. Alo, Y.W. Teh, Mobile and wearable sensors for data-driven health monitoring system: State-of-the-art and future prospect. Expert Syst. Appl. **202**, 117362 (2022)
2. M.S. Yahya et al., LoRa microstrip patch antenna: A comprehensive review. Alex. Eng. J. **103**, 197–221 (2024)
3. P. Clegg, L. Bruciatelli, F. Domingos, R.R. Jones, M. De Donatis, R.W. Wilson, Digital geological mapping with tablet PC and PDA: A comparison. Comput. Geosci. **32**(10), 1682–1698 (2006)
4. R. Uddin, I. Koo, Real-time remote patient monitoring: A review of biosensors integrated with multi-hop IoT systems via cloud connectivity. Appl. Sci. **14**(5), 1876 (2024)
5. U. Nissanov, G. Singh, *Antenna Technology for Terahertz Wireless Communication* (Springer, 2023)
6. J.M. Bangeles, E.M. Tapawan, D.E. Arboleda, Wearable antennas in medical applications: A comprehensive review of design, performance, safety, and clinical applications. Int. J. Res. Publ. Rev. **5**, 709–723 (2024)
7. M. Hassan et al., Significance of flexible substrates for wearable and implantable devices: Recent advances and perspectives. Adv. Mater. Technol. **7**(3), 2100773 (2022)
8. S. Seneviratne et al., A survey of wearable devices and challenges. IEEE Commun Surv Tutor **19**(4), 2573–2620 (2017)
9. S.N. Mahmood et al., Recent advances in wearable antenna technologies: A review. Prog. Electromagn. Res. B **89**, 1–27 (2020)
10. A.J. Johansson, *Wireless Communication with Medical Implants: Antennas and Propagation* (2004)

11. A. Vasjanov, V. Barzdenas, A review of advanced CMOS RF power amplifier architecture trends for low power 5G wireless networks. Electronics **7**(11), 271 (2018)
12. K.N. Bocan, E. Sejdić, Adaptive transcutaneous power transfer to implantable devices: A state of the art review. Sensors **16**(3), 393 (2016)
13. R. Karim, A. Iftikhar, B. Ijaz, I.B. Mabrouk, The potentials, challenges, and future directions of on-chip-antennas for emerging wireless applications—A comprehensive survey. IEEE Access **7**, 173897–173934 (2019)
14. H. Singh et al., Wideband textile multiple-input-multiple-output antenna for industrial, scientific and medical (ISM)/wearable applications. Int. J. RF Microw. Comput. Aided Eng. **30**(12), e22451 (2020)
15. V. Kulkarni, K. Narayana, S.K. Sahoo, A survey on interference avoiding methods for wireless sensor networks working in the 2.4 GHz frequency band. J. Eng. Sci Technol. Rev. **13**(3) (2020)
16. G.A. Casula, G. Montisci, G. Muntoni, A novel design for dual-band wearable textile eighth-mode SIW antennas. IEEE Access **11**, 11555–11569 (2023)
17. M. El Bakkali, N.E.A. El Idrissi, F.E. Tubbal, G.S. Gaba, Optimum design of a tri-band MPA with parasitic elements for CubeSat communications using genetic algorithm, in *Proc. 6th Int. Conf. Wireless Netw. Mobile Commun*, (WINCOM, 2018), pp. 1–4
18. S.K. Sharma, S. Rao, L. Shafai, *Handbook of Reflector Antennas and Feed Systems Vol. I: Theory and Design of Reflectors* (Artech House, 2013)
19. A.Y.I. Ashyap et al., An overview of electromagnetic band-gap integrated wearable antennas. IEEE Access **8**, 7641–7658 (2020)
20. B. Munos et al., Mobile health: The power of wearables, sensors, and apps to transform clinical trials. Ann. N. Y. Acad. Sci. **1375**(1), 3–18 (2016)
21. F. Nikbakhtnasrabadi, Antenna sensing for wearable applications, Ph.D. dissertation, Univ. Glasgow, 2022.
22. F.F. Hashim et al., Key factors in the implementation of wearable antennas for WBNs and ISM applications: A review. Electronics **11**(15), 2470 (2022)
23. B. Wang, Z. Cao, A review of impedance matching techniques in power line communications. Electronics **8**(9), 1022 (2019)
24. P.J. Soh et al., Wearable wireless health monitoring: Current developments, challenges, and future trends. IEEE Microw. Mag. **16**(4), 55–70 (2015)
25. P.J. Soh, G.A.E. Vandenbosch, M. Mercuri, D.M.M.-P. Schreurs, Wearable wireless health monitoring: Current developments, challenges, and future trends. IEEE Microw. Mag. **16**(4), 55–70 (2015)
26. M. Wagih et al., Microwave-enabled wearables: Underpinning technologies, integration platforms, and next-generation roadmap. IEEE J Microw **3**(1), 193–226 (2022)
27. K. Sainath et al., Development of flexible durable multi-slotted antenna for wearable applications. Heliyon **10**(23) (2024)
28. S.F. Abdulkareem, Y.S. Mezaal, Applications of fractal and quasi fractal geometries in slot antenna design: A review. J. Mech. Continua Math. Sci. **14**(4), 216–238 (2019)
29. S. Bukhari, Artificial meaterials for microwave applications, Ph.D. dissertation, Loughborough Univ., 2017.
30. A.T. Mobashsher, A. Abbosh, Three-dimensional folded antenna with ultra-wideband performance, directional radiation and compact size. IET Microw Antennas Propag **8**(3), 171–179 (2014)
31. K.N. Paracha, S.K. Abdul Rahim, P.J. Soh, M. Khalily, Wearable antennas: A review of materials, structures, and innovative features for autonomous communication and sensing. IEEE Access **7**, 56694–56712 (2019)
32. P.S. Hall, Y. Hao, *Antennas and Propagation for Body-Centric Wireless Communications* (Artech House, 2012)
33. A. Ashyap, R. Raad, F. Tubbal, W. Khan, S. Abulgasem, Comprehensive review of wearable antennas with flexible periodic structures for body-effect mitigation. IEEE Access (2025)

34. G. Kaur, A. Kaur, G.K. Toor, Antennas for biomedical applications. Biomed. Eng. Lett. **5**, 203–212 (2015). https://doi.org/10.1007/s13534-015-0193-z
35. A. Reichman et al., Body communications, in *Pervasive Mobile and Ambient Wireless Communications: COST Action 2100*, (2012), pp. 609–660
36. N.H. Hussein et al., A comprehensive survey on vehicular networking: Communications, applications, challenges, and upcoming research directions. IEEE Access **10**, 86127–86180 (2022)
37. J. Liu, H. Liu, Y. Chen, Y. Wang, C. Wang, Wireless sensing for human activity: A survey. IEEE Commun Surv Tutor **22**(3), 1629–1645 (2019)
38. C.R. Ahn et al., Wearable sensing technology applications in construction safety and health. J. Constr. Eng. Manag. **145**(11), 03119007 (2019)
39. M.N. Helmus, D.F. Gibbons, D. Cebon, Biocompatibility: Meeting a key functional requirement of next-generation medical devices. Toxicol. Pathol. **36**(1), 70–80 (2008)
40. M.U.A. Khan et al., Bending analysis of polymer-based flexible antennas for wearable, general IoT applications: A review. Polymers **13**(3), 357 (2021)
41. R. Salvado, C. Loss, R. Gonçalves, P. Pinho, Textile materials for the design of wearable antennas: A survey. Sensors **12**(11), 15841–15857 (2012)
42. S.K. Koul, S. Swapna, G.S. Karthikeya, *Antenna Systems for Modern Wireless Devices* (Springer, 2024)
43. E. E. Atimati, Automated Management of Radio Spectrum and Transmission Power in Heterogeneous Shared Spectrum Networks, Ph.D. dissertation, 2024.
44. S. Geirhofer, J.Z. Sun, L. Tong, B.M. Sadler, Cognitive frequency hopping based on interference prediction: Theory and experimental results. ACM SIGMOBILE Mobile Comput. Commun. Rev **13**(2), 49–61 (2009)
45. Q. Zhao, B.M. Sadler, A survey of dynamic spectrum access. IEEE Signal Process. Mag. **24**(3), 79–89 (2007)
46. L.H. Hemming, *Architectural Electromagnetic Shielding Handbook: A Design and Specification Guide* (Wiley, 2000)
47. S.M. Ali et al., Recent advances of wearable antennas in materials, fabrication methods, designs, and their applications: State-of-the-art. Micromachines **11**(10), 888 (2020)
48. Z. Chen, G. Vandenbosch, X. Zheng, *Wearable Antenna and Rectenna Designs for WBAN and H-IoT Applications* (2024)
49. S.G. Kirtania et al., Flexible antennas: A review. Micromachines **11**(9), 847 (2020)
50. M.A. Ullah et al., A review on antenna technologies for ambient RF energy harvesting and wireless power transfer: Designs, challenges and applications. IEEE Access **10**, 17231–17267 (2022)
51. V. Gupta, R. Kumar, Biomedical telemetry antenna innovations: Progress, uses, and prospects for the future. Prog. Electromagn. Res. B **106** (2024)
52. A. Alexiou, M. Haardt, Smart antennas for future wireless systems. EURASIP J. Wirel. Commun. Netw. **2004**(3), 135–153 (2004)
53. P. Salonen, Y. Rahmat-Samii, H. Hurme, M. Kivikoski, Dual-band E-shaped patch wearable textile antenna. IEEE Antennas Wirel. Propag. Lett. **4**, 149–152 (2005)
54. A. Kiourti, K.S. Nikita, A review of implantable patch antennas for biomedical telemetry: Challenges and solutions. IEEE Antennas Propag. Mag. **54**(3), 210–228 (2012)

Chapter 3
Dipole Antennas for Medical Telemetry Systems

Contents

© The Author(s), under exclusive license to Springer Nature Switzerland AG 2026

A. Segun Ayokunle et al., *Biomedical Antenna Technology*,
https://doi.org/10.1007/978-3-031-95268-5_3

3.1 Introduction

Based on the physical structure, the antennas are classified into wire antennas and aperture antennas. The most common and basic wire antenna is the dipole antenna. It consists of two conductive elements: a wire and a rod. Every conductive rod is of length $\lambda/4$. Therefore, ideally, the length of the dipole antenna is $\lambda/2$, where λ is the wavelength of the desired radiation signal [1]. The antenna is referred to as a half-wave dipole antenna. The half-wave is the most common dipole antenna. However, the use of a dipole is not restricted to the half-wave dipole antenna only. The extension of the arms of the half-wave dipole can give full-wave, one and half-wave dipole, and so on [2]. In all cases, the two sections of antennas are connected end-to-end through a feed line. The feeding is at the center of the antenna. The dipole antenna has maximum radiation perpendicular to the length of the antenna. Therefore, a vertically installed dipole antenna produces an omnidirectional pattern in the horizontal plane. On the other hand, the horizontally mounted antenna produces radiation in the vertical plane [3]. Figure 3.1 shows a horizontally mounted dipole antenna.

3.2 Radiation Mechanism of Dipole Antennas

To understand the radiation mechanism of a dipole antenna, it is important to understand how the free space waves are created in the antenna. Consider a short center-feed dipole antenna, as shown in Fig. 3.2. In Fig. 3.2a, the force lines created between the two arms of the dipole antenna are shown. For the time period $T/4$, the lines have traveled outwardly a distance of $\lambda/4$, where λ is the wavelength of the antenna. Here, it is assumed that the dipole antenna creates a total of three force lines. For the next T/4 period, the same three force lines will travel an additional $\lambda/4$ distance, as shown in Fig. 3.2b. The charge density on each side of the conductors will start to diminish, and exactly at half the time period T/2, the charges on the

Fig. 3.1 Horizontally mounted dipole antenna [4]

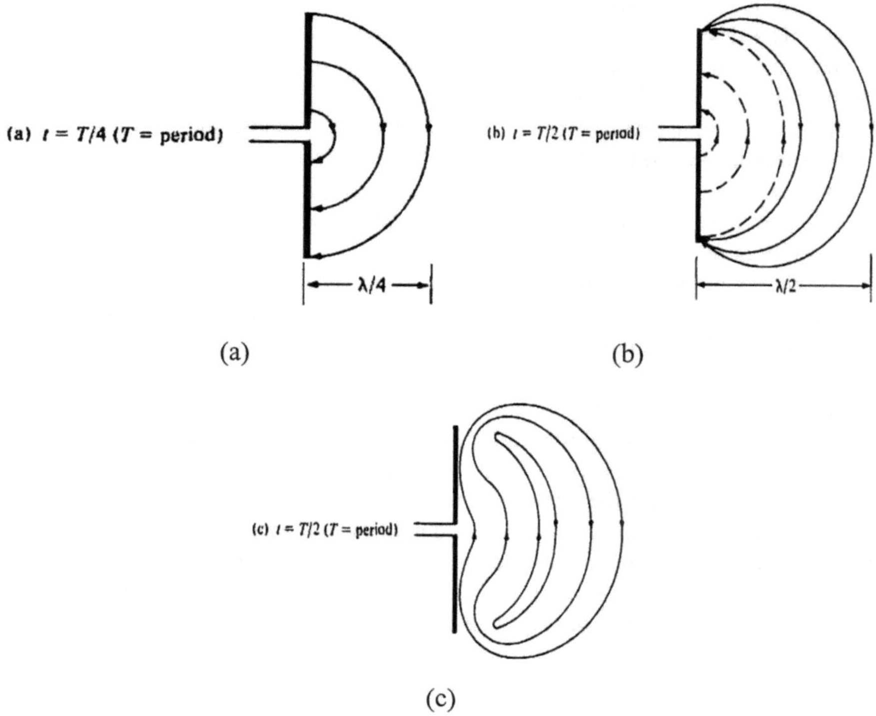

Fig. 3.2 (**a–c**) Radiation mechanism of dipole antenna [5]

conductor become neutralized. The opposite charge on each conductor creates three lines of force. This time, the direction of lines is opposite but travels only $\lambda/4$ radial distance. The charge again starts to diminish. The result is that there are three upward force lines and three downward force lines. As soon as the net charge on antenna conductors becomes "0," the line of forces is detached from the antenna and forms a closed loop. This is shown in Fig. 3.2c. For the next half of the time period, the same process is repeated but in the opposite direction. This whole process will continue infinitely till the dipole antenna gets energy [5].

3.3 Advantages and Limitations of Dipole Antennas in Medical Telemetry

There are many benefits of a dipole antenna that help to get better performance than a conventional wire antenna. Some of the major advantages are as follows:

1. One of the most significant advantages of a dipole antenna is that it is **simple to use**. The design is also simple. Even simple conducting wires can be used to make a dipole antenna. It also has a variable gap size that allows for flexibility in the performance. The dipole antenna is useful in TV transmission.
2. A dipole antenna can be tuned to different frequencies, and it **catches radio frequencies speedily**. This feature makes it a perfect candidate for telemetric medical telemetry applications. These radio frequencies enhance the reception of TV or radio. A better sound and picture quality is achieved at the receiver with the help of dipole antennas.
3. Another benefit of the dipole antenna is that it **receives balanced signals**. High-gain stacked dipole antennas include a two-pole design that helps the antenna receives signals from different frequencies. They also solve the problems of conflicting signals and improve the quality of TV and radio reception. Balanced signals give better quality signals.
4. The dipole antennas are **available in various forms and sizes**, such as basic, folded, and half-wave dipole. Side mount dipole antennas have ends that turn on the backside toward the center to improve signal quality. Other dipole antenna types include the half-wave dipole and half-wave folded dipole antennas.
5. The dipole antenna has an **omnidirectional radiation pattern**. It means that the antenna can send and receive signals from any direction. These antennas get a donut-shaped radiation pattern, and they sit in the middle of the hole. This feature is particularly important when communication is independent of direction. Moreover, it helps to give powerful signals from every direction.
6. As compared to other antennas used for medical telemetry, the dipole antenna gives a **wider frequency range** than other types of antennas. Therefore, it is suitable for applications in MICS, WMTS, and ISM bands.

3.3.1 Limitations of Dipole Antennas

Though a dipole antenna has many advantages, it also has some major limitations.

1. The two-pole design of the antenna **creates a hassle in movement**. Therefore, the use of dipole antennae in implant and wearable applications becomes limited. The printed dipole can be used in such cases. Sometimes, even handling the non-implanted antenna is difficult.
2. A dipole antenna usually operates at **low frequencies**. The operation at low frequency will increase the wavelength and, therefore, the dimension of a dipole antenna. Due to the long wavelength, the powerful signal cannot be received efficiently. Quarter-wave dipole antennas are preferred to improve performance.
3. The **omnidirectional radiation** pattern of the dipole antenna can increase energy consumption. The energy consumption of the antenna becomes very crucial in battery-operated implant devices.

4. As the antenna length reduces, the **efficiency of the antenna** falls. Capacitive reactance will cause an impedance mismatch of the dipole antenna.
5. Along with these, some other limitations are **SAR level, radiation efficiency, miniaturization, biocompatibility**, etc.

3.3.2 Frequencies and Potential Interference

A dipole antenna can be used for implant communication if it falls in the Medical Implant Communication Service (MICS) band between 402 and 405 MHz. The band is mainly used for short-range biomedical communication. However, the band has now been renamed Medical Device Radio Communications (MedRadio), and the range is expanded to 401–406 MHz. The MICS band is limited to short-range communication and also has a limited data rate. Another major application of dipole antennas is in medical telemetry. The FCC has proposed the Wireless Medical Telemetry Service (WMTS) band. In WMTS, a higher data rate is achieved compared to the MICS band. In WMTS, the frequency bands are recommended in the range from 433.1 to 434.8 MHz, 608 to 614 MHz, 868 to 868.6 MHz, and 902.8 to 928 MHz. In some cases, the Industrial Scientific and Medical (ISM) band at 2.4 GHz is also used for communication [6–8] (Table 3.1).

3.3.3 Length of the Dipole, Resonant Frequency, and Radiation Efficiency

The design of a dipole antenna for healthcare applications specifically depends upon whether it is going to be used as an implant, wearable, or telemetry device. Inside an IMD, the area occupied by the antenna is a crucial factor. Different miniaturization schemes are followed for implant antennas. The dipole arms of the antenna are folded. Therefore, the current flow path is enlengthened while reducing the resonant frequency. Due to folded arms, the inductance of the antenna is increased, and impedance matching is enhanced. The successive L-type sections are added to the antenna to achieve proper impedance matching. Increasing the L-type section improves the impedance matching, but miniaturization deteriorates

Table 3.1 Wireless medical telemetry service band

Frequency band	Frequency range
MICS (Medical Implant Communication Services) or Med Radio Band	401–406 MHz
WMTS (Wireless Medical Telemetry Service)	608–614 MHz, 1395–1400 MHz, 1427–1432 MHz
ISM band	2.4–2.5 GHz

efficiency significantly. On the other hand, the off-body antennas are relaxed to have a folded structure. In telemetry applications, different standard dipole antennas, such as quarter-wave dipole, half-wave dipole, and full-wave dipole, can be implemented. In medical telemetry, these conventional antennas are best suited for use at the base stations for patient monitoring purposes. In the case of wearable and WBAN antennas, the dipole antenna should be woven or knitted into the clothes, and therefore, the folded antenna should not be used here. Furthermore, the sturdiness of the antenna is important. The antenna should have sufficient strength so that the body part's movement will not interrupt the communication [9].

For the selection of resonant frequency, consider three different dipole antennas: half-wave dipole, full-wave dipole, and three-half-wave dipole. The corresponding current flows are shown in Fig. 3.3.

3.3.3.1 Resonant Frequency

From the above figures, it is clear that a discrete set of wavelengths follows the equation:

$$L = n * \left(\frac{\lambda}{2}\right) \tag{3.1}$$

It means that a discrete set of frequencies can follow the equation:

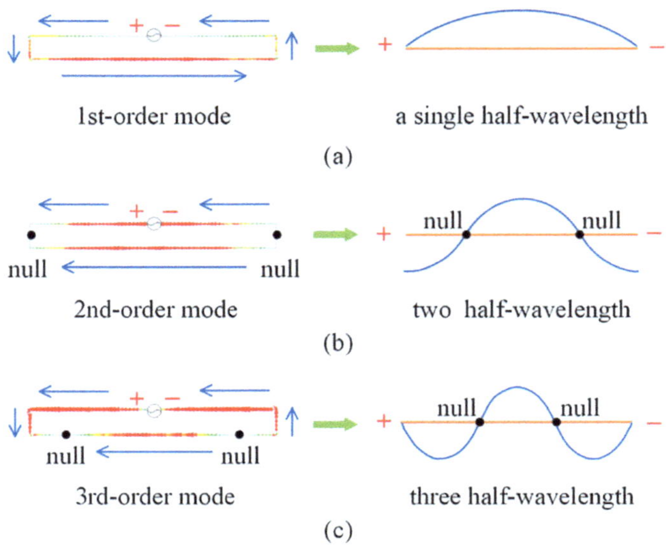

Fig. 3.3 Schematic of the first-order, second-order, and third-order modes of the linear dipole Antenna [6]

$$f = n * \left(\frac{c}{2L} \right) \tag{3.2}$$

Here, the c is the speed of light.

Therefore, the possible resonant frequencies for the dipole antennas are

$$f = \left(\frac{c}{2L} \right), 2 * \left(\frac{c}{2L} \right), 3 * \left(\frac{c}{2L} \right) \tag{3.3}$$

3.3.3.2 Radiation Efficiency

For an antenna, the parameter to know how well it radiates or receives power is known as radiation efficiency. It is calculated as the ratio of powers. The ratio of radiated power and total power is referred to as radiation efficiency for the transmitter antenna. Since the radiated power is dependent on the radiation resistance, the formula for calculating the radiation efficiency is as follows:

$$\text{Radiation efficiency} \left(\eta \right) = \frac{R_{\text{rad}}}{R_{\text{rad} + R_{\text{loss}}}} \tag{3.4}$$

$$\text{Radiation resistance } R_{\text{rad}} = \frac{80\pi^2 \left(\frac{l}{\lambda} \right)^2}{\Omega} \tag{3.5}$$

In the above equation, R_{rad} represents radiation resistance, and R_{loss} is the resistance used to calculate the power loss. Further, l is the length of the antenna, and λ is the operating wavelength. It can be observed from the above formulas that the radiation efficiency can be increased by reducing the power loss and increasing the radiating power. In dipole antennas, the radiation resistance has to be much larger and closer to the characteristic impedance, so the efficiency should reach close to 100% [7, 10].

3.3.4 Generation of Electromagnetic Waves and Radiation Pattern

The process of electromagnetic wave generation in a dipole antenna is governed by the current that passes through each conducting wire. In the process, the incoming wave's electric field pushes the electrons in the wire back and forth, and the polarity of the two wires alternates. The motion of the electron generates the current in the wires. The direction of the current changes alternately. The current flow in the wires

Table 3.2 Dipole antennas and corresponding radiation patterns

Dipole antenna name	Design of antenna	Radiation pattern
Half-wavelength dipole antenna (l = $\lambda/2$)	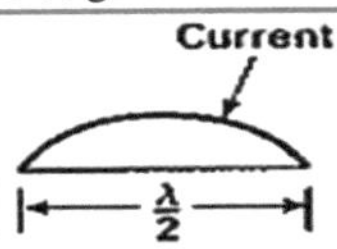	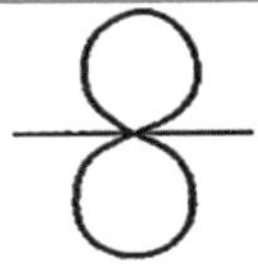
Full-wave dipole (l = λ)	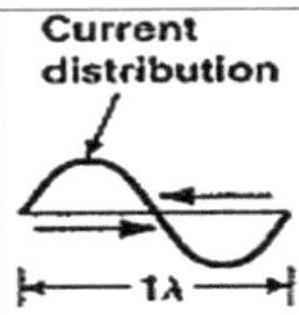	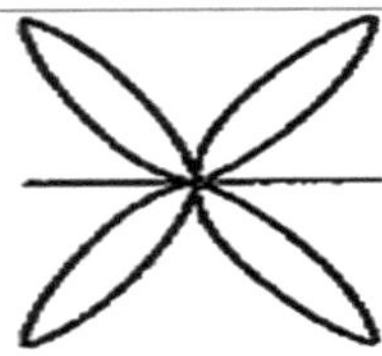
(l = $3\lambda/2$)	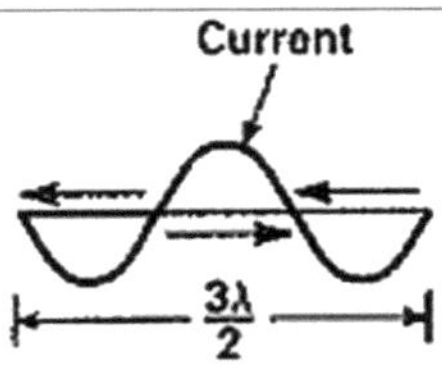	
(l = 3λ)	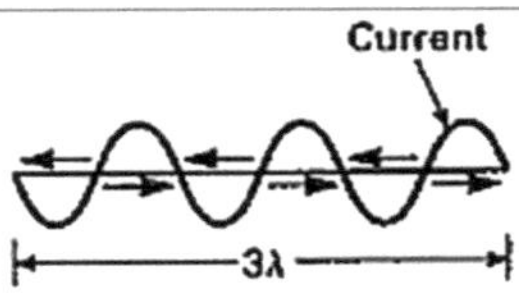	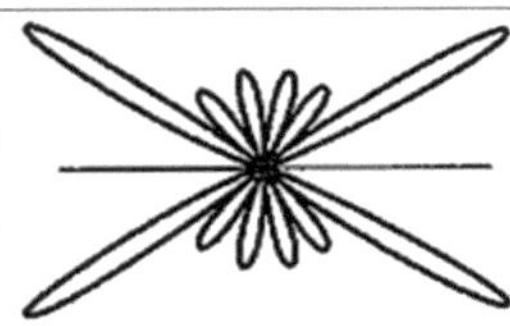

will generate a magnetic field around the dipole antenna. The generation of the magnetic field is given by Biot–Savart law.

$$B = \frac{\mu_{0NI}}{2R} \tag{3.6}$$

Here, B = magnetic field intensity, μ_0 = permeability of free space, N = number of turns, I = current intensity, and R = radius. These electric and magnetic fields are combined to form a radiating EM field. Precisely, the frequency of the generated EM field is the same as the frequency of the alternating current (Table 3.2).

3.4 Dipole Antennas for Implantable Medical Devices

The use of dipole antennas in implantable medical devices is moderately popular. The IMD can support various diagnostic and therapeutic functions. There are some severe challenges, such as dimension, biocompatibility of the device, radiation pattern, transmission range, data rate, SAR, etc., in the use of dipole antennas as IMD.

3.4.1 Use of Dipole Antennas in Implantable Medical Devices

The human body implant devices are placed inside the body to improve the lives of patients suffering from chronic disease. The antenna implanted inside the body will regularly transmit the data for monitoring. Whenever a patient reaches the hospital for a checkup, the implant status is observed there. The recorded data by the implant antenna is transmitted to the receiving base station. RF technology is useful even if the patient is waiting in the queue or coming for regular checkups. In some cases, the home care unit is placed to monitor the status at home, and the patient does not need to visit the hospital regularly [11].

The main purpose of any implantable medical device (IMD) is to perform various diagnostic and therapeutic functions like drug delivery, sensing, and stimulation. The antennas are integrated with the implanted devices to receive signals from the control base available outside the body and wirelessly transmit vital in-body fluctuations like glucose, heartbeat, body temperature, and blood pressure.

3.4.2 Challenges of Implantable Antennas

Mostly, the implanted antennas are mounted over the conductive case of IMD. The characteristics of the antenna can vary because of this conductive case. The effect can be vital in near-electric field intensities. The near-electric field intensity increases. The near magnetic field remains unchanged. Since the IMD is implanted in biological tissue, the antenna parameters, such as radiation efficiency and radiation pattern, depend on the tissue's electrical properties (Table 3.3).

To ensure patient safety, the specific absorption rate (SAR) should be maintained in a specified range. Furthermore, the asymmetry of the human body makes it very challenging to implant the antenna. The tissue size is not the same throughout the body [12].

3.4.3 Constraints and Biocompatibility

Biocompatibility is one of the major challenges in IMD design. The biocompatibility of the implant antenna with the tissue environment is very necessary. The conductive nature of human tissues can short-circuit the metallization of the implanted

Table 3.3 Biomedical tissue permittivity, conductivity, and loss tangent at different frequencies

	Relative permittivity			Conductivity (S/m)			Loss tangent		
f (MHz)	Fat	Muscle	Skin	Fat	Muscle	Skin	Fat	Muscle	Skin
403	5.57	57.10	46.71	0.004	0.797	0.689	0.328	0.622	0.658
915	5.46	54.99	41.33	0.051	0.948	0.872	0.185	0.339	0.414
2400	5.28	52.73	38.00	0.104	1.739	1.464	0.145	0.242	0.283
5800	4.95	48.49	35.11	0.293	4.962	3.717	0.183	0.317	0.328

antenna. To achieve biocompatibility, a coating is applied over the antenna. So, a dielectric separation is done between the antenna and the tissue. The Rogers substrate used in the antenna is replaced by Alumina, a ceramic material. Polydimethylsiloxane (PDMS) can also be used to design implant antennas. The implant antennas are tested by immersing them in phantom liquid, so the electrical properties of human tissues are imitated. Miniaturization is another constraint in implant antennas. The dimension of antennas should be small to reduce the wavelength effectively. Since the operating frequency is too small, achieving this small wavelength is challenging.

3.4.4 Blockage

The dipole antennas have omnidirectional radiation patterns. It means it radiates power equally in one of the planes, i.e., azimuthal, elevation, etc. The omnidirectional pattern is best suited for telemetry applications. For an implantable device, the antenna should be directional so that it can be implanted in such a way that the antenna radiation has to penetrate the thin part of the tissue. The dipole antenna radiation cannot penetrate the thick obstacles, and the antenna suffers from a blockage problem. The maximum radiation should come out of the body, and reliable communication with a sufficient data rate will be set up. To avoid the blockage of radiation, the thickness of different tissues must be considered. Therefore, for the purpose of analysis, the tissue models are created with a thickness of 2.5 mm, representing implantation beneath the skin, and 25 mm is considered for the muscles and bones (Fig. 3.4).

3.4.5 Diffraction

The bending of RF around the obstacle is referred to as diffraction. In the human body, the electrical properties vary from tissue to tissue. The electrical properties of different tissues are given in the table above. The diffraction may be severe if the antenna is implanted deeper into the body. The stomach, muscles, and intestines can affect the performance of the implant antenna. Further, sometimes, the resonant frequency of the antenna can also vary.

3.4.6 Scattering and Reflection

The low-frequency communication system causes challenges not only to IMD but also to the control unit established outside the human body. The sensitivity of the receiver, scattering due to close objects, and unwanted reflections of EM waves from human organs are major reasons for the path propagation loss.

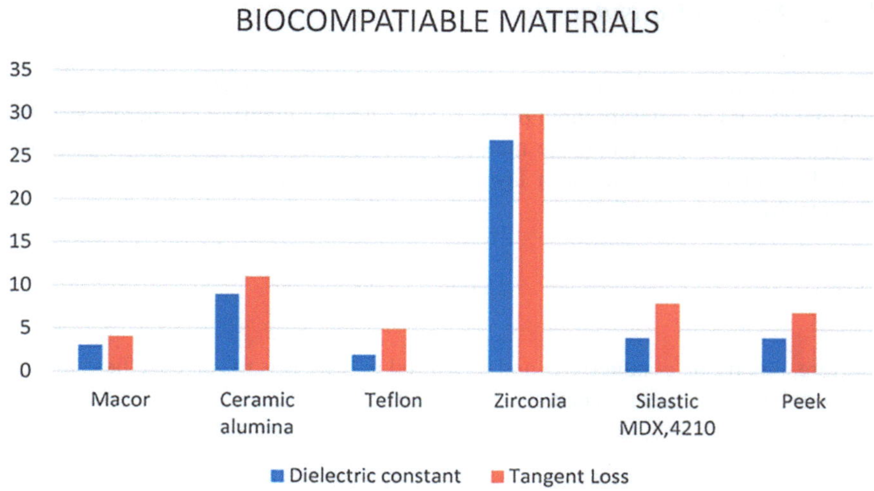

Fig. 3.4 Graphical representation of biocompatible materials and their dielectric constant and tangent loss

3.5 Dipole Antennas in Wearable Health Monitoring

The advancements in research in military and healthcare applications are making wearable technology very popular. The wires of the dipole are woven or knitted into the threads of the cloth to implement the antenna. The cellulose filter paper can also be used as a dielectric substrate to implement printed dipoles. The electric properties ε_r are dependent on the effective thickness of the paper and can be varied from application to application. Natural rubber and elastomeric polydimethylsiloxane (PDMS) are other dielectric materials that can also be used to make printed dipole antennas. The permittivity of these materials is variable and can be modified by mixing some other materials with them. The substrate with a low dielectric constant is mostly used to reduce the surface wave and get a higher impedance bandwidth. The implementation of a dipole antenna on a rigid substrate, polytetrafluoroethylene (PTFE), or tectonic ceramic can also be used [13].

3.5.1 Integration of dipole Antennas

A smooth integration of the antenna in cloth is always desirable. For that purpose, weaving, knitting, or pasting can be implemented. The flexible antennas are integrated into wearable technology. The dipole antennas are fabricated using flexible material. To meet the requirements of users, the wearable dipole antenna should be flexible, lightweight, compact, durable, and comfortable. Non-stretchable denim fabric made with cotton yarn is preferably used for wearable antennas.

3.5.2 Design Considerations

In the simulation, the results should be checked for a 10 g average specific absorption rate (SAR) at different input powers. The SAR is a measure of energy absorbed by body tissues. SAR has to be measured for the wearable antenna. The antenna should be placed near the skin layer. The formula used for SAR calculation is given below.

$$\mathrm{SAR} = \sigma \frac{|E|^2}{\rho} \tag{3.7}$$

Here, σ is tissue conductivity and measured in S/m, E is the electric field in V/m, and ρ in Kg/m^3 is the mass density of the tissue.

3.6 Dipole Antennas for In-Body Communication

The in-body communication, where one implant communicates with another implant, is gaining popularity nowadays. In-body propagation demands that radio waves penetrate different human tissues. The frequency-dependence and variation in the thickness of the tissue make communication very difficult. Further, the complexity is increased as no experimental analysis is feasible inside the human body. The experimental setup requires the placement of probes and cables to collect data, which cannot be done in the case of implant-to-implant communication. In-body communication has not been explored too much. Therefore, no reliable channel model is available to analyze the communication setup. The human body has tissues of varying sizes with variable electromagnetic properties. Therefore, the channel model should incorporate these changes correspondingly to the position of the implant in the body. In-body communication leads to a communication setup among implant devices. Since the implants do have a limited power source in the form of a battery, the processing capability of transmitting and receiving data is very limited. Therefore, an on-body processor is usually placed, as shown in Fig. 3.5. Generally, the implant devices need to send the impulses to the other implants. For this purpose, two methods, ultrasonic wave transmission and high-frequency EM wave transmission, are used.

The FCC assigned the MICS (401–405 MHz) band for implant devices. The MICS is an ultra-low-power and unlicensed frequency band. The band is mainly designed for diagnostic and therapeutic functions associated with implanted medical devices such as pacemakers and defibrillators. This MICS frequency provides better propagation characteristics for IMDs. Further, at this frequency, an antenna with reasonable dimensions can be designed. To enable the intrabody communication among IMDs, the signal generation capacity is very limited. The IMDs are battery-driven and cannot generate a very high-frequency signal for

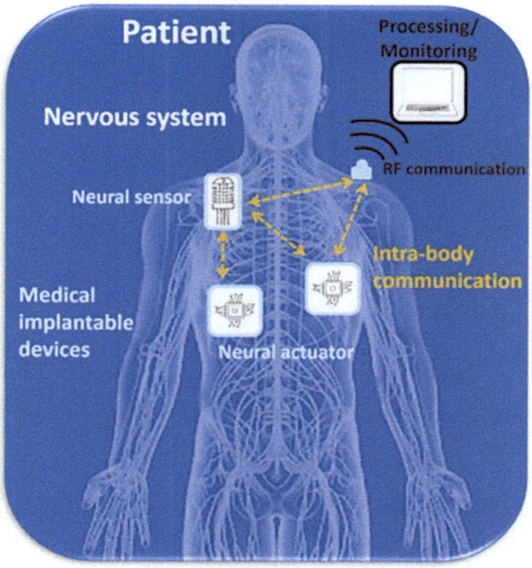

Fig. 3.5 In-body communication in medical devices and implants [14]

communication. Therefore, two approaches are prominently used. The two approaches are shown in Fig. 3.6. First is the use of ultrasonic signals for communication. The human body has tissues with 65–70% water. It is observed in many studies that the frequency ranges between 700 KHz and 5 MHz can achieve optimum performance in the condition of attenuation and directivity. However, ultrasonic communication suffers from fading caused by reflection and refraction from the high-density tissues. The limited communication range and diffusive nature of communication are the major challenges in the use of ultrasonic waves for in-body communication [15].

The second approach includes a high-frequency EM signal transmission. The coverage distance, tissue penetration, and fading behavior depend on the EM signal frequency. The frequencies up to 2 GHz can be used, but a trade-off between the battery power and the frequency of use is needed. Although the high-frequency offers a large bandwidth of up to 2 MHz at an energy cost, the large bandwidth exhibits good communication properties with a high data rate. Sometimes, a parallel plate waveguide is formed by using two conductors on both sides of the fat tissue. The contrast in dielectric properties between the skin and the fat and the muscle and fat confined the signal in the fat layer of tissue. The tissue behaves like a low-loss parallel plate waveguide to transmit the signal over a longer distance inside the body. The method looks worthy even for microwave communication [14].

The human body is not suitable for the transmission of EM signals. Therefore, for in-body communication, it is important to understand the effect of the RF signals when they pass through the tissues of the human body. The various tissues have different dielectric constants, impedances, and conductivities. These parameters are also dependent on the frequency of in-body communication. The human body

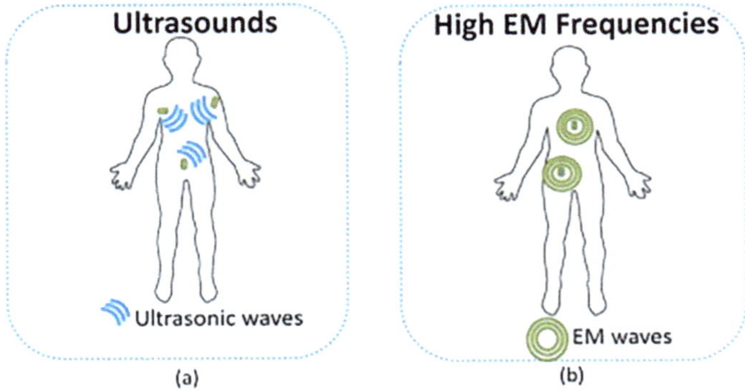

Fig. 3.6 Communication within the human body: (**a**) ultrasonic wave and (**b**) high EM frequencies [14]

causes significant absorption losses and radiation pattern loss. The absorption by a tissue is dependent on both frequency and the water and ionic content within the tissue. The water and ionic content alter the field characteristics of the tissue. Further, the liquid nature of the human body causes the RF attenuation. On the other hand, the skeletal structure inside the human body causes diffraction and refraction of the RF signal. The different body components offer varying degrees of RF interactions. The selection of the communication wavelengths is crucial in this regard. The large wavelength signal has to penetrate the large cross-section of the human body. Meanwhile, the small wavelength signal, i.e., high-frequency signal, suffers from a scattering effect. Therefore, a proper characterization of the body is required to develop a model for communication. Also, the tissue-related properties and selection of operating frequency become crucial.

3.7 Dipole Antennas for Wireless Medical Imaging

There are many methods of imaging, but the medical imaging techniques where antennas are used profoundly are broadly categorized among.

1. Magnetic resonance imaging
2. Microwave imaging
3. Ultrasound imaging
4. Terahertz imaging

Medical imaging is one of the fields where dipole antennas are used widely. The dipole antenna is preferred because of its simplicity, adaptability, and effective use for both the transmission and reception of RF signals. The significant areas of imaging where the dipole antenna can be used include magnetic resonance imaging (MRI), microwave imaging, ultrasound imaging, and terahertz imaging [16].

The MRI is a rapidly developing non-invasive technique used for clinical investigation of various fatal diseases such as brain tumors, early detection of breast cancer, etc. Nowadays, ultrahigh-field MRIs are gaining popularity, where a magnetic field of up to 7 T is used for diagnostics. This high field is provided by superconducting magnets. A better SNR performance with high resolution is achieved. For MRI applications, the dipole antenna is used as an RF coil to transmit or receive signals from the body. However, the high-field MRI system design has significant hardware development and safety challenges. For example, when we use 298 MHz at 7 T, the high permittivity of the human body tissues reduces the wavelength by approximately 13 cm in the abdominal cavity. The conductivity of tissue may cause a reduction in penetration capability. Further, the SAR levels also reached harmful levels [17].

Microwave imaging is preferred for breast cancer detection. A prototype of the microwave head imaging is shown in Fig. 3.7. The dipole antenna operates at microwave frequencies (1–10 GHz) and interacts with diseased tissue. This frequency is particularly suitable for near-field applications. The microwave frequencies can penetrate the dielectric materials easily and are useful for detection, diagnostics, classification, and monitoring. It is used in brain stroke diagnosis, breast cancer detection, and continuous blood glucose monitoring. Further, microwave imaging is used in cutting-edge applications like knee injuries, torso scanning, thermal therapy, etc. Portability and cost-effectiveness are the main advantages of microwave imaging devices [16, 18].

The high-resolution imaging is part of ultrasound imaging. Here, only the modified dipole antennas can be implemented for EM-acoustic interactions. Dipole antenna application in terahertz imaging is underutilized and limited to surface-level imaging, and it mainly deals with tissue burns.

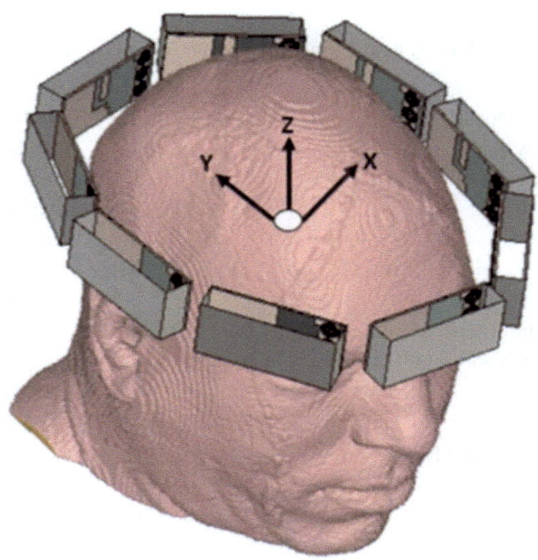

Fig. 3.7 A prototype of microwave head imaging system [16]

3.8 Frequency Band Selection for Medical Telemetry Dipole Antennas

3.8.1 Frequency Range Selection

The Federal Communications Commission has specified the frequency ranges for the medical telemetry applications. Largely, three bands are preferred for medical telemetry applications (Table 3.4).

3.8.2 Applications and Regulatory Restrictions

The FCC standards are followed worldwide. In Europe, 868–870 MHz and 2.4 GHz are also used as ISM bands for medical telemetry applications. The U. S. Food & Drug Administration (FDA) has provided the regulations for the application-specific use and restrictions in medical telemetry. The medical devices should strictly operate in the WMTS frequency bands to reduce electromagnetic interference. The details of WMTS bands are given in the table above. The FDA also put a limitation on the transmit output power and out-band emission for the protection of other applications. The user has to get permission to transmit more than 80 km (50 miles) for some bands and 32 km (20 miles) for some other bands. This frequency coordinator maintains the record to avoid conflicts with frequency reuse. The power used in the devices is also limited to 1 watt EIRP. The FCC has also implemented regulations for electromagnetic compatibility to avoid interference from nearby devices. For electromagnetic compatibility, the standard IEC 60601-1-2 is also followed. For

Table 3.4 Federal Communications Commission specified medical telemetry bands

Bands (as per FCC)	Frequency ranges	Applications
Wireless medical telemetry service (WMTS)	608–614 MHz	FCC allocated channel TV 37 extensively for medical telemetry applications
	1395–1400 MHz	A shared channel is used in telemetry to minimize interferences
	1427–1432 MHz	A shared channel is used for communication with the other telemetry equipment
Industrial scientific and medical Bands (ISM)	902–928 MHz	Used for non-critical medical telemetry
	2.4–2.4835 GHz	Used for general applications like Wi-fi, Bluetooth, etc.
	5.8 GHz	Used for high data rate applications
VHF and UHF band	174–216 MHz	Limited use because of broadcasting rights. The band used TV channels 7–13
	470–608 MHz	Previously used, but now the use is limited, the band used TV channels 14–36

the purpose of telemetry data safeguarding, HIPAA (Health Insurance Portability and Accountability Act) restrictions are followed [19].

3.8.3 Advantages of Broadband dipole Antennas

The broadband dipole antenna is handy for operating over a wide range of frequencies to ensure multiple applications. A single broadband antenna can replace the use of multiple narrowband antennas. This is a kind of future-proofing where the same wideband antenna can be used for future applications. The broadband dipole antennas are best suited for diverse applications like patient rooms, ambulances, etc. Further, the multiband functionality can help mitigate the interference by changing the operating frequency. To reduce signal loss and ensure reliable monitoring for a stable healthcare facility, the broadband dipole antenna provides a cost-efficient solution.

3.9 Problem Formulation

Telemetry communication in healthcare applications is the need of the hour. The dipole antenna is among the preferred antennas for wireless medical telemetric communication. Therefore, a dipole antenna can radiate in the WMTS band. One of the main objectives of this chapter is to design a dipole antenna. The antenna should fall in the WMTS band for communication. In the WMTS band, the dipole antenna can be used as a wearable device, infusion pumps, remote patient monitoring, cardiac monitors, and smart hospital beds. The use of a dipole antenna in wireless medical telemetry can also support monitoring multiple patients simultaneously in large hospitals. The size of the antenna, electric length, resonance frequency, transmission range, directivity, SNR performance, etc., are the significant challenges in the dipole antenna design that must be taken care of. Therefore, the antenna should:

- Operate within the frequency ranges suitable for **WMTS bands**.
- Ensure stable performance despite changes in the **surrounding environment**.
- Achieve sufficient **bandwidth**, **efficiency**, and **gain**.

3.10 Related Work

The design of dipole antennas for implant and telemetry communication in medical applications has been extensively explored, focusing on achieving compactness, biocompatibility, and suppressed interference and noise operation across medical communication bands. Key studies include the literature survey mentioned in Table 3.5.

3.11 Simulation Result and Discussion

The dipole antenna is designed on Roger RT/Duroid 6010 dielectric substrate. The antenna covers an area of 20×80 mm^2, and the length of the dipole is 70 mm. The power to the dipole antenna is fed by the microstrip line. The length and width of the microstrip line are 12×3.8 mm^2. The antenna consists of a partial ground plane, and the length of the ground plane is 9.0 mm. The proposed planner dipole antenna is shown in Fig. 3.8. The antenna dimensions are shown in Table 3.6. The antenna resonates at 1.3 GHz; the return loss of the antenna is given in Fig. 3.9. The resonant behavior of the antenna can be understood with the help of the current distribution, as shown in Fig. 3.10.

The dipole antenna resonates at 1.3 GHz and covers the bandwidth of 1.3–1.42 GHz. The bandwidth of the antenna is calculated at -10 dB. The antenna return loss is shown in Fig. 3.9. The current distribution of the antenna is shown in Fig. 3.10. With the help of current distribution, we can understand the radiation behavior of the antenna. The antenna provides a bi-directional radiation pattern in the E plane and an omnidirectional radiation pattern in the H plane. The radiation pattern of the antenna is shown in Fig. 3.11. The radiation behavior of the antenna can further understand with the 3D radiation diagram. The 3D radiation pattern is shown in Fig. 3.12.

The next few portions of the chapter are dedicated to advanced applications of the dipole antenna. The application covered the use of a dipole antenna array for directional telemetry. The chapter further covers the different methods to reduce noise and interference to improve SNR. Finally, the open research challenges for future research are mentioned in the chapter.

3.12 Dipole Arrays for Directional Medical Telemetry

Medical telemetry requires focusing the RF energy for transmission and reception. The beam's focus in a particular direction is achieved by using the dipole array. The dipole array has many advantages in the design of medical telemetry communication systems. The advantages, such as improved directivity, improved SNR performance, improved signal strength, and interference mitigation, are the most common. The array of dipole antennas can also provide the beam-steering capability to the telemetry system. Making a dipole array is a challenging task since the system designer has to take care of the number of dipole elements, spacing between the elements, orientation, feeding, and polarization of the elements [31]. The size of the array, i.e., the number of elements in the array, is set to achieve the desired gain.

Table 3.5 Literature survey of antennas for biomedical applications

Ref. no.	Key feature	Details
[3]	Wearable, triband dipole antenna is designed with low SAR	The paper introduced a novel design of a wearable E-shaped, tri-dipole antenna. The antenna is printed on a 2 mm thick denim fabric of $\varepsilon_r = 1.7$. By putting in a rectangular patch with L-shaped cuts, the bandwidth at higher frequencies is improved. The antenna performance is observed in free space and as well as at different parts of the body. The 10 g average SAR at different input powers (maximum 250 mW) is observed to understand the effect of the antenna on the human body without spacing or shielding
[20]	A graphene nanoflake-printed dipole antenna is designed for short-range RFID and sensing application	In the paper, a graphene nanoflake-printed antenna is designed. The conductive ink has been printed on a paper substrate to achieve a conductivity of 0.43×10^5 S/m. The meandered-line dipole antenna is fabricated as a proof of concept. The observed maximum gain is −4 dBi, and the −10-dB bandwidth ranges from 984 to 1052 MHz (6.67%). The observed radiation efficiency is 32%. The antenna has applications in mid-band short-range RFID and sensing applications. Because of the use of the screen-printing technique, mass production of antennas is possible
[21]	A wearable textile dipole antenna is designed for application in the body area network	This paper addresses the design of a textile dipole antenna based on embroidery techniques. The antenna operates at the 2.45 GHz ISM band for WBAN applications. Two medium stitch density embroidery patterns, satin fill and contour fill, are used. The results are obtained in terms of return loss, radiation pattern, realized gain, and efficiency
[22]	A textile diamond dipole antenna and artificial magnetic conductor are designed for wearable application	The antenna is designed and tested under wearable and body-centric conditions. The antenna is entirely made of textiles, which means both substrate and conducting parts are made with textiles. Directive patterns with high gain are obtained, hence minimizing the radiation toward the human body. The body-centric measurements are done and particularly investigate bending, wetness, and SAR. A notable reduction in SAR is achieved with the introduction of the artificial magnetic conductor sheet
[23]	An implantable planar dipole antenna for wireless MedRadio band and application in biotelemetry is designed	An implantable planar dipole antenna MedRadio band is proposed. Initially, a skin-implantable antenna model is developed, and then a prototype is fabricated. The dipole antenna exhibits a physical volume of approximately 18.1 mm ^{3}. Effective size is reduced by optimally canceling its high capacitance through L-shaped reactive loading. The antenna demonstrates reasonable gain, radiation pattern shape, and SAR. A liquid skin medium is used to validate the performance of the antenna

(continued)

Table 3.5 (continued)

Ref. no.	Key feature	Details
[24]	Wideband planner loop-dipole antenna is designed for point-to-point communication	In the paper, a planar wideband balanced composite antenna is designed. The antenna consists of a symmetrical dipole and an antipodal loop radiator. The parameters such as return loss, radiation pattern, and gain of the antenna are obtained. The antenna offers an impedance bandwidth of 129% between 3.26 and 15 GHz. A stable end-fire radiation patterns with good polarization purity are obtained. The proposed antenna has an application in point-to-point communication
[25]	A multiband slot dipole antenna was designed with CPW feed	In this paper, total three antennas, two multiband antennas, and one broadband antenna are investigated. The antennas use slot stepped-impedance resonators (SIRs) and coplanar waveguide (CPW) feed. All three antennas are fabricated on 1-mm FR4 substrates . Compared to the uniform-impedance resonator (UIR), the SIR shows a better control over spurious response. The broadband antenna is designed by allocating the three resonant frequencies close enough to achieve a 98.7% fractional bandwidth with a return loss of more than 6 dB between 1.5 and 4.4 GHz
[26]	A miniature planar dipole antenna designed for implant devices	A miniature planar dipole antenna is designed for integration into wireless biomedical devices. The proposed exhibits a compact size suitable for implant application and offers the resonance in the MICS frequency band. The analysis of the antenna is conducted inside a homogeneous 2/3 muscle-equivalent phantom. The designed antenna is robust, allowing placement in different positions and demonstrating sufficient bandwidth at a very small package. The SAR level is also in a limit
[27]	Designed an implantable folded dipole antenna for in-body wireless communication	This paper presents an implanted folded antenna of very compact size. The antenna is designed to operate at in UHF bands (0.951–0.956 GHz). For the performance evaluation, an equivalent human phantom arm is considered. The proposed antenna offers a maximum antenna gain of −23.5 dBi and for viable wireless communication
[28]	Designed a printed folded dipole antenna for tissue implant application	In this paper, a tissue-implantable folded dipole antenna of very compact size is designed. The antenna operates in the MICS frequency band. The antenna design and analysis are done in the flat phantom with electrical properties equivalent to 2/3 of human muscle tissue. The antenna is printed on a high-dielectric substrate layer and exhibits a frequency resonance at 402 MHz with a wide 10-dB impedance bandwidth of 105 MHz. The antenna radiates an omnidirectional far-field pattern with a maximum gain of −31.2 dB. The SAR range is also satisfactory

(continued)

Table 3.5 (continued)

Ref. no.	Key feature	Details
[29]	Designed a coaxial dipole antenna, aiming at treatment of brain tumor	In the paper, the authors investigated a coaxial dipole antenna that is aimed to have applications in medical treatment for brain tumors. The antenna used to heat the tumor part only by generating a localized heating pattern. The heating pattern is controlled in the longitudinal direction of the antenna by using array. To control the heating in the axial direction, the authors introduced a coaxial dipole antenna
[30]	Designed a folded dipole antenna with unidirectional radiation and low impulse distortion for head imaging applications	The paper presented a compact 3D antenna for microwave-based head imaging systems. The antenna is fed by a coplanar waveguide. It consists of a slot-loaded folded dipole structure with four furled sides. The proposed antenna offers low-distorted radiation losses with an impulse fidelity factor of more than 80% toward the boresight direction in both near and far fields. The antenna has a compact size of 0.29 × 0.08. A 67% fractional bandwidth over 1.1–2.2 GHz is achieved. Further, a 16-element array of the antenna is tested. Also, the antenna array is employed on a human head model to detect and locate a hemorrhagic brain injury
[12]	Designed an implanted dipole antenna for an RFID-based patient monitoring system	In this paper, a tiny dipole antenna for an RFID tag will be implanted into the body. The antenna is encapsulated by a silica glass to reduce the radiation effects on the human body. The antenna is simulated in a human arm model at 924 MHz. The antenna is also tested in free space and silica glass capsule. The antenna is relatively very small dimension compared to the band wavelength with good performances in S parameter, sufficient bandwidth, and omnidirectional radiation pattern
[11]	Designed a folded dipole antenna for medical implants at 900 MHz	The paper proposes a folded dipole antenna for an implantable device to monitoring of patient wirelessly. The device is assumed to be implanted through a syringe injection. The antenna is operated in UHF band 924 MHz, which is band of Indonesian frequency allocation for RFID applications. The antenna offers good performance of S-parameters, impedance bandwidth, radiation pattern, and gain. The antenna offers gain for more than 10 m range communication with 250 MHz bandwidth (VSWR ≤1.5)

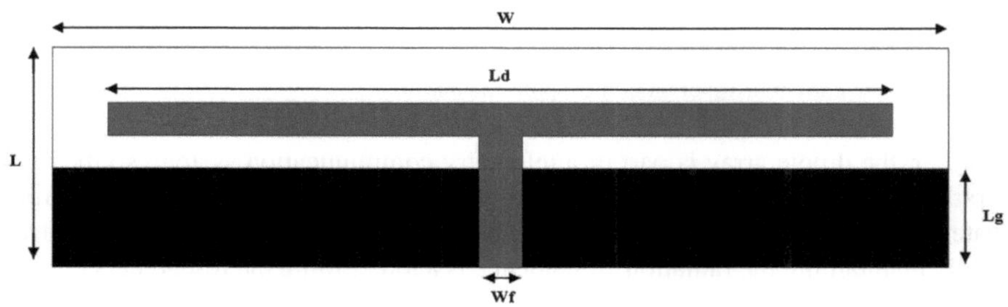

Fig. 3.8 Proposed dipole antenna

Table 3.6 The dimension of the proposed dipole antenna

Parameters	Dimensions (mm)	Parameters	Dimensions (mm)
L	20.0	Ld	70.0
W	80.0	Wf	3.8
Lg	9.0	Lf	12.0

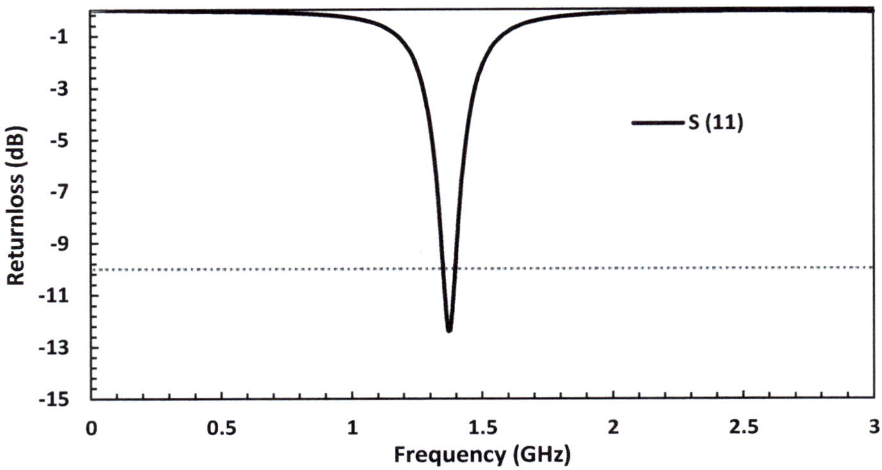

Fig. 3.9 Return loss of the proposed dipole antenna

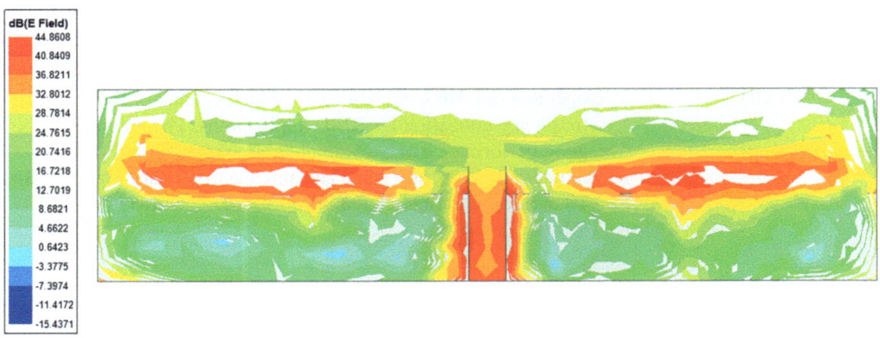

Fig. 3.10 The current distribution of the dipole antenna

Further, the dipole array is part of a telemetry communication system, so the integration with the other components of the system is also a crucial parameter to mitigate the interference. The optimal energy transfer between the elements of the array while maintaining the radiation coincidence is another thing the researchers have to care about. An example of a dipole antenna array is shown in Fig. 3.13.

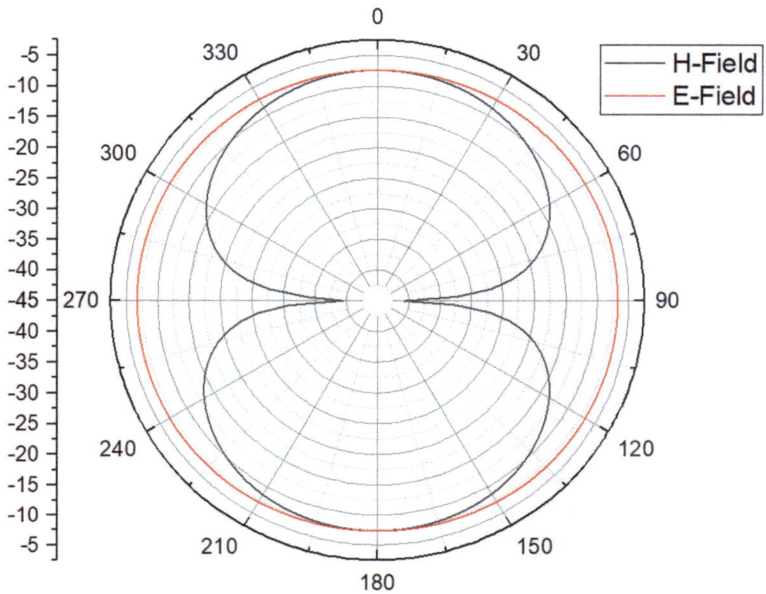

Fig. 3.11 2D radiation pattern (E and H Plane) of the dipole antenna

One of the significant healthcare technology requirements is monitoring a remote patient. Generally, a health monitoring system includes sensors on the patient's body, a monitoring module installed near the patient, i.e., on the patient's body or at the bed unit, an antenna, a processor, and a display connected to it. Sometimes, the processor and the display device are merged and replaced by a computer. The conventional health monitoring system is shown in Fig. 3.14. In a patient health monitoring system, the antenna plays a pivotal role if the processor and displaying device are installed far from the patient and a wireless transmission is required. The use of an antenna becomes crucial when the health monitoring system is installed inside a hospital, and the health officials have to monitor multiple patients simultaneously from one place. In such a scenario, every antenna installed at a different bed should focus its signal on a distant receiver. A directive antenna is desired to achieve a better focus of EM waves. The directivity of the antenna is enhanced if an array of antennas is used. The conventional dipole antennas have an omnidirectional radiation pattern. To achieve directive communication, it is observed that a dipole array is a better approach compared to a simple dipole antenna. Therefore, telemetric communication prefers the antenna array.

The dipole array is extensively used for directional medical telemetry. The arrays can focus the signal to or from a specific patient. In most cases, the patient is considered to be lying in a fixed position in a hospital bed or home care, and the dipole arrays are used at both transmitter and receiver ends to achieve greater focus of RF signals. The directional array of the dipole antenna can be part of the critical care

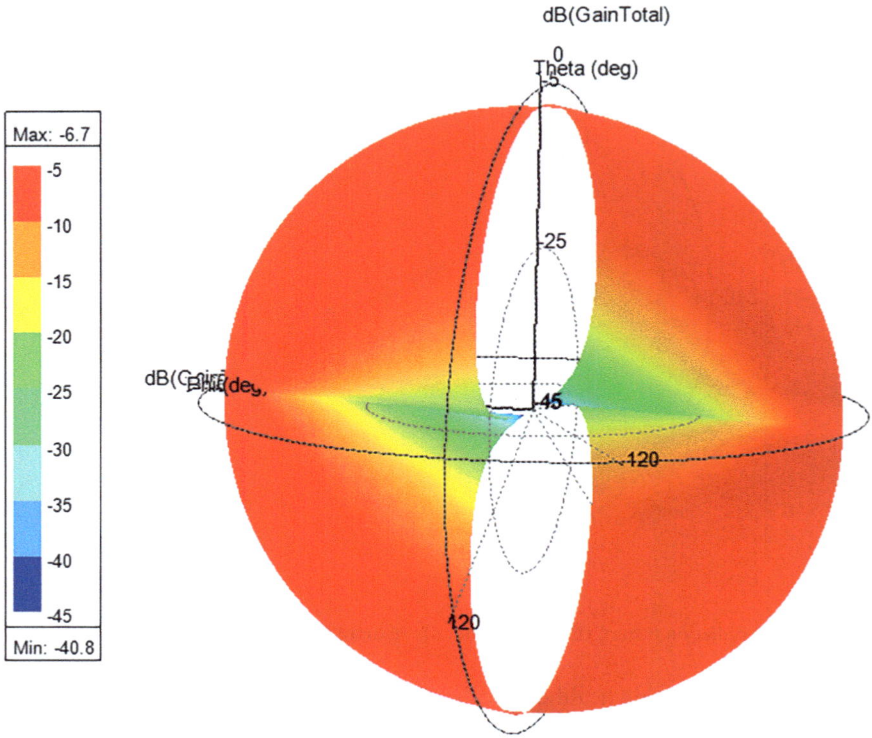

Fig. 3.12 3D radiation pattern of the dipole antenna

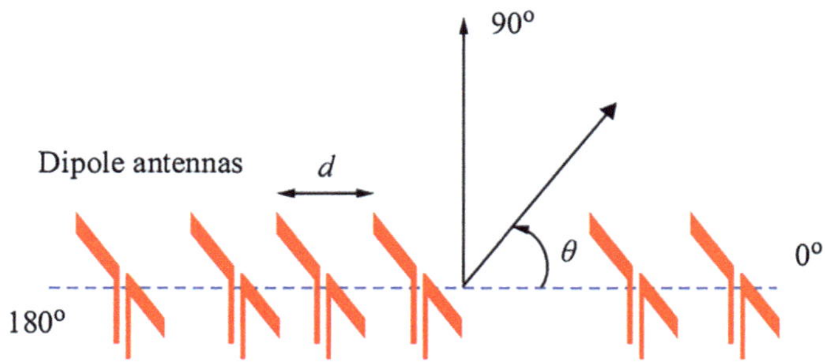

Fig. 3.13 Dipole antenna array

unit installed near the patient's bed, and then it transmits signals to the distant base station with advanced processing capability. The setup is particularly useful for processing the data in a hospital with multiple beds, such as an ICU and a single base station unit. The health care officials can monitor all the patients simultaneously through the base station unit. Further extension of the use of directional dipole

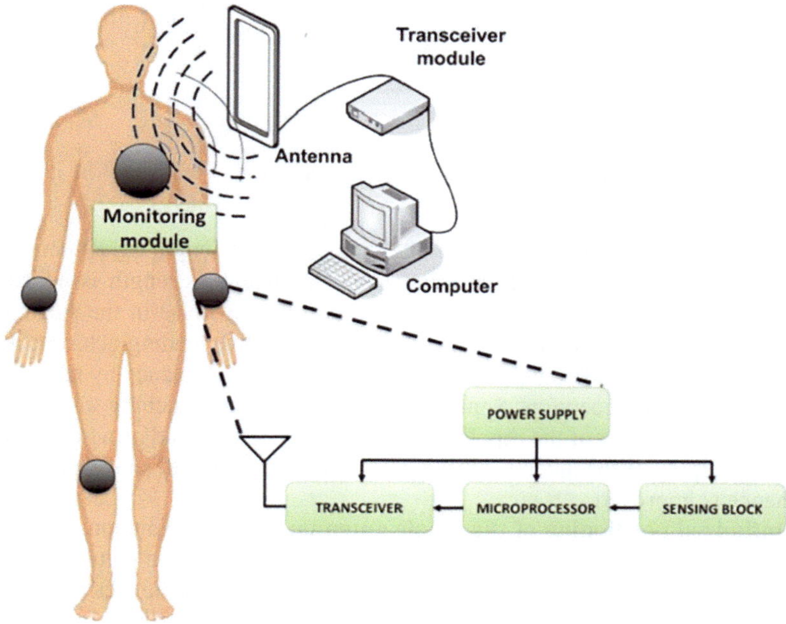

Fig. 3.14 Wireless health monitoring system [13]

arrays in medical telemetry is in the ambulance and emergency response. A stable communication link is ensured between the moving ambulance and the hospital by using a dipole array. Though the idea of using a dipole array looks fascinating, the complexity of design, size constraints, and beam misalignments are the major challenges in the design of the dipole array.

3.13 Enhancing Signal-to-Noise Ratio in Medical Telemetry Dipole Antennas

The telemetry data transmission with a dipole antenna suffers from poor SNR performance. The accuracy and reliability of the physiological data transmission can also be compromised sometimes. Different schemes have been implemented to improve the SNR characteristics of medical telemetry dipole antennas. Some of them are mentioned in the subsequent section.

$$\mathrm{SNR} = \frac{P_{\mathrm{signal}}}{P_{\mathrm{noise}}} = \left(\frac{A_{\mathrm{signal}}}{A_{\mathrm{noise}}}\right)^2 \tag{3.8}$$

$$\mathrm{SNR}(dB) = 10^{10}\log\left(\frac{P_{\mathrm{signal}}}{P_{\mathrm{noise}}}\right) = 20^{10}\log\left(\frac{A_{\mathrm{signal}}}{A_{\mathrm{noise}}}\right) \tag{3.9}$$

3.13.1 Techniques for Noise Management

1. **Optimizing the antenna design**: The dipole antenna length adjustment can be done to change the resonance frequency. The antenna's length is adjusted to show resonance at a frequency that suffers from less noise in the surrounding environment. Further, tapering the element or resistive loading achieves a wide bandwidth. The wide bandwidth helps to enhance performance by reducing the chances of mis-reception. Sometimes, the impedance matching circuit is also attached at the end of the dipole to reduce the reflections and provide maximum power supply to the antenna. The signal power is increased, so the SNR performance is improved.
2. **Optimized positioning of the dipole antenna:** The dipole antenna is position-sensitive. The dipole antenna has an omnidirectional radiation pattern. To get the radiation in a particular plane, the dipole antenna must be placed precisely. Further, the positioning of the dipole antenna is crucial for body area network applications. The antenna should get minimum attenuation from the body tissue. At the time of installation, the polarization of the dipole antenna should also be checked. The polarization of the transmitting and receiving antennas should be matched for a convenient telemetric communication link.
3. **Reducing the environmental interference:** In medical telemetry, some frequency bands are shared by other applications. These frequency bands suffer from the noise caused by the equipment installed for other applications. The external noise is reduced by shielding around the dipole antenna. The shielding of the operating frequency can also be done by creating multiple parallel channels in a shared band.
4. **Using high-quality substrate:** Sometimes, the losses in dipole antennas are also caused by using poor-quality materials as the substrate. Materials like copper and silver will be the preferred option to reduce the conductive losses. The substrate material should be selected to reduce the absorption of RF energy.
5. **Improving the directivity of antenna:** The directive antennas are less exposed to interference. The dipole antenna array can be used for this purpose. The beamforming is another technique that uses a phased array of dipole antennas. The use of metamaterials can also enhance the directivity and efficiency of the antenna.

Many other approaches can be implemented to reduce the noise in the telemetric communication system. However, these approaches are not restricted to use with dipole antennas only. The approaches are as follows.

Using low-noise amplifiers: In the telemetry communication system, along with the antenna, the LNAs are also implemented to amplify the weak signals.

Employing diversity/MIMO antenna techniques: One of the main advantages of diversity is that it can reduce interference. The spatial diversity is implemented to mitigate the multipath fading. In some cases, the polarization and frequency diversity are also helpful in reducing the interference.

Improving system integration: The medical telemetry communication system includes an antenna, modulator, amplifier, and sometimes batteries for a power supply. Integrating these components with fewer cables and connector requirements will improve the system's noise performance. The proper grounding should also reduce the noise in the system.

Improvement in signal processing capability: The signal processing capability can also improve the noise performance of the telemetry communication systems. With advanced signal processing tools, noise filtering becomes more effective. Also, error correction algorithms are implemented to reconstruct the signals.

3.14 Open Research Challenges

The dipole antenna in telemetry applications offers many advantages, such as simplicity in use, flexibility in design, omnidirectional radiation pattern, availability of different standard antennas, and ease of arranging as an array. The dipole antennas also have some challenges related to their design and application in telemetric communication. Some of the open research challenges that need to be addressed by the researchers are mentioned below.

1. Miniaturization of dipole antennas: Dipole antenna use is extended to wearable technology, where antenna size is crucial. In wearable telemetry, the antennas are woven or knitted into the cloth, so the shape and size of the dipole antenna should support the weaving. The metamaterials can also be used to reduce the size of the dipole antenna without altering the performance. A novel feeding method with minimum space requirement is also desired for such wearable telemetric dipole antennas.
2. Wideband and multiband operations: For medical telemetry operations, multiple frequency bands are suggested by the FCC. Out of all bands, the WMTS bands are very prominent. The ISM band and some VHF and UHF channels can also be used in telemetry applications. The wideband dipole antenna that extends its operation across all the telemetry bands will be the preferred choice for the receiver. On the other hand, if a multiband dipole antenna has resonances in these multiple telemetric bands, the antenna can also be very useful. Such a multiband antenna will reduce the hardware requirement in telemetric communication. The optimized impedance matching in the wideband antenna is further a challenge.
3. Beamforming and beam steering: A pivotal challenge with using dipole antennas in medical telemetry applications is to achieve directionality and beamforming. The beamforming and directionality will be useful in monitoring multiple

patients simultaneously from distant locations. The use of compact and adaptive arrays of dipoles can be a solution. A software-defined beam-steering mechanism can also be developed.

4. Interference and noise reduction: Most medical telemetric devices are placed in crowded environments where other equipment shares the same frequency band. This will lead to potential interference and noise. Therefore, a robust design for the dipole antennas is desired to suppress the noise or interference. Using noise-cancellation filters can be an option, but in this case, a trade-off has been made with the antenna size. A frequency hopping technique can also be implemented to suppress the interference.

Along with all the above-mentioned challenges, some general issues related to dipole antennas and telemetry systems need to be addressed. It includes the scalability of antennas in the arrays, integration with other components of the communication such as modulator, LNAs, integration with wearable and implant devices, long-term reliability, energy efficiency, cost-effective design, use of substrate or the elements to make dipole components, biocompatibility, safety in the use, signal loss, tissue interaction, indoor and outdoor testing of the communication setup, data security and privacy, and high-performance imaging. Researchers can explore these fields to achieve better medical telemetric communication with the help of dipole antennas.

3.14.1 Summary

The presented chapter explores the basics of the dipole antenna by providing information on its radiation mechanism, advantages, and challenges. To discuss the radiation mechanism, the steps of how an opposite polarity on the poles of the dipole antenna generates the EM waves and how this EM field is detached from the antenna are covered. Further, the chapter explores the advantages and limitations of dipole antennae in medical applications, particularly telemetry applications. The chapter also provides details on frequency bands that can be used for medical telemetry applications and gives details about the selection of the length of the dipole elements, the resonance mechanism, and the radiation efficiency. The next topic covers the application of dipoles in medical implants. The details of MICS bands and their key features, limitations, and challenges are mentioned in the topic. Then the chapter's focus swiftly moved to the telemetry application of the dipole antenna. In this section, the frequency ranges for WMTS bands and details of the different regulatory restrictions imposed by the FCC and FDA are discussed. The portion also gives the use of a broadband dipole antenna in medical telemetry. In the same section, a dipole antenna design for medical telemetry applications is given. The antenna resonates at 1.4 GHz. The resonant frequency falls in the WMTS band, and hence, the antenna is suitable for the wireless medical telemetry application.

References

1. R. Barzegari, C. Ghobadi, J. Nourinia, A dual-band dipole array antenna with fan-beam radiation for biomedical and telecommunications applications. Results Eng. **24**, 103136 (2024)
2. M. Li, K.M. Luk, Wideband magneto-electric dipole antennas, in *Handbook of Antenna Technologies*, ed. by Z. Chen, (Springer, Singapore, 2015. [Online].). https://doi.org/10.1007/978-981-4560-75-7_49-1
3. H.I. Azeez, H.-C. Yang, W.-S. Chen, Wearable triband E-shaped dipole antenna with low SAR for IoT applications. Electronics **8**(6), 665 (2019. [Online].). https://doi.org/10.3390/electronics8060665
4. Vertical Dipole Antenna. IW5EDI [Online]. Available: https://www.iw5edi.com/ham-radio/2975/vertical-dipole-antenna
5. C.A. Balanis, *Antenna Theory: Analysis and Design*, 2nd edn. (Wiley, New York, 1997)
6. W. Hu et al., Compact wideband folded dipole antenna with multi-resonant modes. IEEE Trans. Antennas Propag. **67**(11), 6789–6799 (2019). https://doi.org/10.1109/TAP.2019.2925188
7. A. Khan, S. Bashir, and F. Ullah, Electromagnetic bandgap wearable dipole antenna with low specific absorption rate. Proc. 2018 Int. Conf. Comput., Math. Eng. Technol. (iCoMET), Sukkur, Pakistan, 2018, pp. 1–4, doi: https://doi.org/10.1109/ICOMET.2018.8346441.
8. Medical implant communications service (MICS) Federal Register. Rules Reg. **64**(240), 69926–69934 (1999)
9. S. Ahdi Rezaeieh, K.S. Bialkowski, A. Zamani, A.M. Abbosh, Loop-dipole composite antenna for wideband microwave-based medical diagnostic systems with verification on pulmonary edema detection. IEEE Antennas Wireless Propag. Lett. **15**, 838–841 (2016). https://doi.org/10.1109/LAWP.2015.2476515
10. N. M. Borkar and P. K. Parlewar, Antennas in medical applications. Proc. IEEE Indian Conf. Antennas Propag. (InCAP), Jaipur, India, 2021, pp. 152–154, doi: https://doi.org/10.1109/InCAP52216.2021.9726335.
11. Basari, D. C. Sirait, F. Y. Zulkifli, and E. T. Rahardjo, Simple folded dipole antenna for medical implant communications at 900 MHz band. Proc. IEEE Asia Pacific Microw. Conf. (APMC), Kaohsiung, Taiwan, 2012.
12. D. C. Sirait, Basari, F. Y. Zulkifli, and E. T. Rahardjo, An implanted dipole antenna for RFID-based patient monitoring system. Proc. Int. Conf. QiR, Yogyakarta, Indonesia, 2013, pp. 142–145, doi: https://doi.org/10.1109/QiR.2013.6632553.
13. T. H. Nguyen, T. L. H. Nguyen, and T. P. Vuong, Printed wearable dual band antenna for remote healthcare monitoring device. Proc. IEEE-RIVF Int. Conf. Comput. Commun. Technol., 2019, pp. 1–5.
14. A. Vizziello, M. Magarini, P. Savazzi, and L. Galluccio, Intra-body communications for nervous system applications: Current technologies and future directions, Comput. Netw., vol. 227, 2023 Art. no. 109718.
15. Federal Communications Commission, Washington, DC, USA, Federal Communications Commission. 2012. [Online]. Available: http://www.fcc.gov
16. M.S. Islam, M.T. Islam, A.F. Almutairi, A portable non-invasive microwave-based head imaging system using compact metamaterial loaded 3D unidirectional antenna for stroke detection. Sci. Rep. **12**, 8895 (2022). https://doi.org/10.1038/s41598-022-12860-8
17. R. Balafendiev et al., An antenna based on three coupled dipoles with minimized E-field for ultra-high-field MRI. IEEE Trans. Antennas Propag. **70**(10), 9083–9092 (2022). https://doi.org/10.1109/TAP.2022.3195515
18. C. Origlia et al., Review of microwave near-field sensing and imaging devices in medical applications. Sensors **24**, 4515 (2024). https://doi.org/10.3390/s24144515
19. U.S. FDA, Wireless Medical Telemetry Systems [Online]. Available: https://www.fda.gov/medical-devices/wireless-medical-devices/wireless-medical-telemetry-systems

20. T. Leng et al., Graphene Nanoflakes printed flexible meandered-line dipole antenna on paper substrate for low-cost RFID and sensing applications. IEEE Antennas Wireless Propag. Lett. **15**, 1565–1568 (2016). https://doi.org/10.1109/LAWP.2016.2518746
21. I. Gil, R. Fernández-García, J.A. Tornero, Embroidery manufacturing techniques for textile dipole antenna applied to wireless body area network. Text. Res. J. **89**(8), 1573–1581 (2019)
22. K. Kamardin et al., Textile diamond dipole and artificial magnetic conductor performance under bending, wetness and specific absorption rate measurements. Radioengineering **24**(3), 729–738 (2015)
23. S. Bakogianni, S. Koulouridis, An implantable planar dipole antenna for wireless MedRadio-band biotelemetry devices. IEEE Antennas Wireless Propag. Lett. **15**, 234–237 (2016). https://doi.org/10.1109/LAWP.2015.2439039
24. W.-J. Lu, W.-H. Zhang, K.F. Tong, H.-B. Zhu, Planar wideband loop-dipole composite antenna. IEEE Trans. Antennas Propag. **62**(4), 2275–2279 (2014). https://doi.org/10.1109/TAP.2014.2299820
25. S.-W. Chen, D.-Y. Wang, W.-H. Tu, Dual-band/tri-band/broadband CPW-fed stepped-impedance slot dipole antennas. IEEE Trans. Antennas Propag. **62**(1), 485–490 (2014). https://doi.org/10.1109/TAP.2013.2287523
26. S. Bakogianni and S. Koulouridis, Performance of a novel miniature antenna implanted into the human trunk for medical telemetry applications. Proc. 9th Eur. Conf. Antennas Propag. (EuCAP), Lisbon, Portugal, 2015, pp. 1–5.
27. H. Lin, M. Takahashi, K. Saito, K. Ito, Performance of implantable folded dipole antenna for in-body wireless communication. IEEE Trans. Antennas Propag. **61**(3), 1363–1370 (2013). https://doi.org/10.1109/TAP.2012.2227099
28. S. Bakogianni and S. Koulouridis, Design of a novel compact printed folded dipole antenna for biomedical applications. Proc. 8th Eur. Conf. Antennas Propag. (EuCAP), The Hague, Netherlands, 2014, pp. 3178–3182, doi: https://doi.org/10.1109/EuCAP.2014.6902503.
29. S. Kikuchi, K. Saito, M. Takahashi, and K. Ito, "Control of heating pattern for interstitial microwave hyperthermia by a coaxial-dipole antenna - aiming at treatment of brain tumor," Electron. Commun. Jpn I Commun., 2007. [Online]. Available: https://doi.org/10.1002/ecja.20390
30. A.T. Mobashsher, A.M. Abbosh, Compact 3-D slot-loaded folded dipole antenna with unidirectional radiation and low impulse distortion for head imaging applications. IEEE Trans. Antennas Propag. **64**(7), 3245–3250 (2016). https://doi.org/10.1109/TAP.2016.2560909
31. U.S. Patent US8203497B2, Dual polarized dipole wearable antenna. Jun. 2012.

Chapter 4
Helical Antennas for Wireless Body Area Networks (WBAN)

Contents

© The Author(s), under exclusive license to Springer Nature Switzerland AG 2026
A. Segun Ayokunle et al., *Biomedical Antenna Technology*,
https://doi.org/10.1007/978-3-031-95268-5_4

4.1 Introduction

A prominent application of wireless communication devices is in body area networks. It comprises devices such as sensors, actuators, and transceivers implanted inside the body, placed on the surface of the body, or digested in the body. The fast advancements in the field of Internet of Things (IoT) have made WBANs an intriguing topic for research across a wide range of applications. Remote health and fitness monitoring, rehabilitation, military and sports training, active combat operations, livestock farming, interactive gaming, personal information sharing, secure authentication, assisted living, etc., are just a few of the many uses for WBAN in IoT applications. WBAN is becoming an increasingly cost-effective solution in many industries that require IoT applications. Furthermore, without a wireless connection, it would be impossible to gather the volume of data generated by WBAN. Hence, the use of an antenna is very desirable in the WBAN system. A great challenge in WBAN is the power management. A key requirement is to consume the minimum energy to run sensors. Power consumption for intra-WBAN and inter-WBAN data transmission is high, and selecting the appropriate antenna for a given application is also crucial. Further, the antenna design has a significant impact on throughput, channel loss, and power consumption. A variety of antennas, such as dipole, monopole, helical, PIFA, and loop antennas, are used for different WBAN applications. Most of the time, low-profile antennas are preferred for WBAN applications [1].

4.1.1 WBAN Architecture

For a particular application, different research groups offer a variety of WBAN architectures. These architectures are made to optimize performance, and communication protocols suit the architecture for specific applications. The three-tier WBAN architecture is the most common architecture, but it is also possible to develop different architectures based on body-to-body (B2B) communication, machine-to-machine (M2M) communication, and ad hoc architectures. The three-tier architecture is shown in Fig. 4.1. In the figure, tier 1 includes the different sensors placed on or inside the human body. It includes electroencephalography (EEG) sensors, electromyography (EMG) sensors, electrocardiography (ECG) sensors, and peripheral oxygen saturation (SPO2) sensors. There are three possible ways to place the sensors: they can be attached to the body, implanted in the body, or swallowed. The tier is also referred to as intra-WBAN, as the communication happens among the different WBANs. Tier 2 is used to have a communication gateway. The gateway includes a Wi-Fi router, mobile phone, etc. The gateway must have Internet connectivity to proceed with the collected data further. At this tier, it is possible that one WBAN communicates with another WBAN, and hence the tier is also referred to as the inter-WBAN communication tier. The communication among different gateways and end users is categorized as tier 3. This tier is also known as beyond the

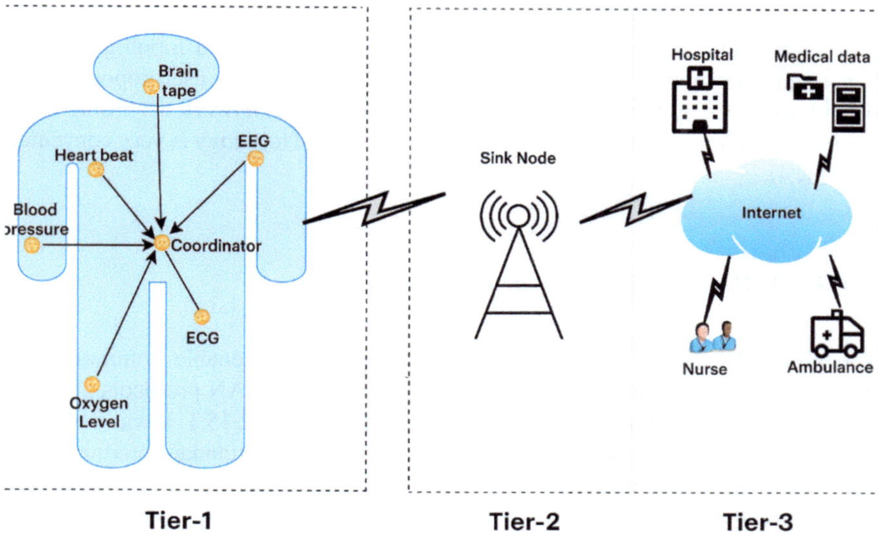

Fig. 4.1 A 3-tier WBAN Architecture [2]

WBAN tier. Cloud computing can be used at either tier 2 or tier 3. It is also possible to use cloud computing between tier 2 and tier 3 [2, 3].

4.1.2 WBAN Topologies

There are four common topologies preferred in the WBAN architecture. The first one is to start topology. The star topology uses one centered access point. The communication is governed by the access point. All the sensors have equal authority to reach the access point. The preferred use of the topology is when the sensor nodes do not need to communicate with each other. They simply want to communicate the access point to deliver the collected data. In some cases, if a sensor node wants to communicate with another sensor node, it is done through the access point. The access point is the mediator for communication between the sensor nodes. In contrast, the mesh topology provides a communication link among all the nodes available at the same level. Every sensor can directly communicate with the other without permission from the access point. Individual communication of sensor nodes is permissible here. The feature looks effective but sometimes causes delays in the data transmission. The full mesh topology is less preferred until the application demands it. The third topology used in WBAN is the tree topology. Here, the priority levels for the nodes are defined. The central node means that the access point concept is available in tree topology as well. The highest-priority sensors can directly communicate with the access point. The remaining nodes have to communicate with the access point through hops. Sensor-to-sensor communication in a tree topology always needs more than one hop unless it is between parent and child nodes. The fourth and last topology makes use of all the abovementioned topologies and is hence named a hybrid topology. This is the most commonly used topology as it provides the liberty to the WBAN designers to mold the network according to the application. The addition of the new sensor nodes in the topology is very convenient here [1, 4].

4.1.3 WBAN Communication Protocol

The protocols are assets of the rules and regulations to enable communication between different entities. The IEEE has offered many WBAN protocols and standards. The first breakthrough standard is coined as IEEE 802.15.1. It was developed in 2002 for Wireless Personal Area Network (WPAN). The standard predominantly uses Bluetooth technology for communication. The standard defined the physical layer protocols for communication in WPAN. Another standard, IEEE 802.15.4, with low-rate WPAN, was introduced in 2003. The standard specifically focuses on the medium access control (MAC) layer for longer battery specifications for fixed and mobile WPAN. Later, in 2003, the IEEE 802.15.2 was introduced as a coexisting network. It offers a high data rate WPAN. A couple of amendments have also been made to the standards, but most are geographic region-oriented. The IEEE 802.15.5 standard is provided for interoperable, scalable, and stable WPANs. Furthermore, to cater to the developments in the WBAN applications, a new standard, IEEE 802.15.6, was specifically developed for WBAN. The WBAN standard operates in three categories, namely implant, on the surface, and external. Out of these three categories, the external has the highest communication range, and that is up to 5 m away from the body. The other two categories have very limited transmission ranges and require precise modelling of the channels to cater to the communication between the nodes. There are a few challenges that still stand in the way of protocol development. These challenges are mentioned below. To discuss the challenges, the International Standard Organization-Open System Interconnection (ISO-OSI) based 7 layers or the Transmission Control Protocol-Internet Protocol (TCP-IP) based four-layer architectures are followed for the protocol design [5, 6].

Physical layer design challenges in WBAN:

1. Interoperability and temperature control
2. Topological changes
3. Bandwidth and data rate variations
4. Constant signalling
5. Interference reduction and fault acceptance
6. Quality of service (QoS) and Quality of Experience (QoE)

MAC layer design challenges in WBAN:

1. Dynamic Channel Assignment (DSA)

2. Control and protocol overhead
3. Synchronization and consistency
4. Packet scheduling, error control, error detection and correction, and fault acceptance
5. Data flow control and energy conservation
6. Delay control and security

Network layer design challenges in WBAN:

1. Optimum routing and optimal path selection
2. Network conditioning and real-time streaming
3. Localization and mobility of the sensor nodes
4. Traffic control, congestion control, and throughput optimization
5. End-to-end packet delivery and multipath routing
6. Security, QoS, and fault tolerance

Transport layer design challenges in WBAN

1. Reliable transport of data packets
2. Congestion control
3. Self-configuration
4. Energy awareness
5. Biased implementation
6. Constrained addressing

Application layer design challenges

1. Efficient interfacing with the nodes and other WBANs
2. Congestion and flow control
3. Bandwidth allocation
4. Packet loss recovery
5. Energy efficiency
6. Security

Among all the layers, the physical layer design plays a pivotal role in communication. The physical layer deals with physical parameters such as power allocation, modulation type, bandwidth selection, interference control, etc. The use of an antenna can ensure the efficiency of communication by maintaining energy efficiency, channel selection, and bandwidth optimization. As mentioned above, the WBAN is created through in-body, on-body, or off-body communication. The in-body communicating devices can be further classified as implantable, ingestible, and injectable devices. The use of different in-body WBAN devices is shown in Fig. 4.2. Different antennas are used according to the applications, as mentioned in Table 4.1. This means that we can use certain types of antennas for a specific implant device. This is because of the operating frequency range constraints. The in-body WBAN devices, their preferred antennas, and the operating frequency bands are mentioned in Table 4.1 [7].

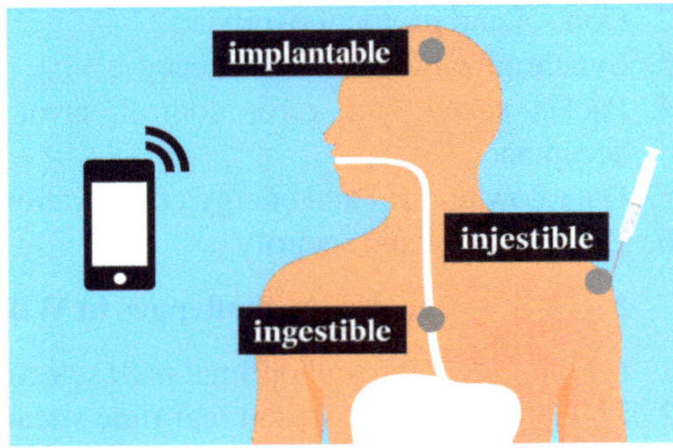

Fig. 4.2 Different types of in-body devices to create a WBAN [7]

Table 4.1 Different types of in-body devices and their preferred antennas

Type of in-body WBAN devices	Preferred antennas	Operating frequencies
Implantable devices	PIFA Patch antennas Loop antennas Monopole antennas	402 MHz (prominently used), 433 MHz, 868 MHz, 915 MHz, 1.4 GHz, 2.45 GHz (ISM band), 6 GHz (UWB band)
Ingestible devices	Helical antennas Spiral antennas PIFA	433 MHz, 500 MHz, 800 MHz, 1.2 GHz, 1.4 GHz and 2.4 GHz
Injectable devices	Loop antenna Dipole antenna	132 KHz, 2 MHz, 13.56 MHz and 915 MHz

The on-body antennas mostly cover the printed antennas and are used in wearable applications. Here, the antennas are printed not only on the conducting materials but also some special cloths are used to print the antennas. For the off-body antennas, the dimension of the antenna is not the constraint, and hence, a variety of antennas, such as a horn antenna, parabolic reflector, helical antenna, helical array, and spiral antenna, can be used to achieve high data rate and to increase the degree of directivity. Wearable antennas are not restricted to tracking and monitoring in medical healthcare, but they can also be useful in military, entertainment, and rescue applications. The uses of the wearable antennas are mentioned in Table 4.2. In the table, the classification of applications for wearable antennas is not restricted to only healthcare; non-medical fields of application are also mentioned [8].

Helical antennas have drawn a lot of interest due to their potential in biomedical applications, especially wireless body area networks (WBAN). For effective wireless communication in difficult in-body situations, these antennas' distinctive benefits—such as their small size, circular polarization, and broadband performance—are

Table 4.2 Application of wearable antennas in medical and non-medical fields

Broader field	Application
Medical and healthcare	Glucose monitoring, breast cancer detection, endoscopy, GPS tracking of patients, wearable thermometer, oximeter, wearable Doppler antenna
Military	The antenna can be used in bulletproof jackets to check the effectiveness of the jacket when a soldier is shot. It can be used to transmit a SoS to a nearby control room and inform them about the current medical status of the military personnel. GPS tracking and fatigue status of the pilots
Rescue	Fire detection, mining operations, etc. are the places where the wearable antenna is particularly useful
Entertainment	Smartwatches, music jackets, intelligent shoes, etc. are some of the uses. Another application in entertainment is in the field of virtual reality and augmented reality
Space and outer space	The wearable antenna can be useful for space and outer space missions

essential. With an emphasis on biocompatibility and miniaturization, this paper examines the properties, uses, and design of helical antennas in the biomedical domain.

4.2 Helical Antenna Characteristics and Advantages

4.2.1 Helical Antenna Key Characteristics

The helical antenna is preferred in biomedical applications such as implantable devices, capsule endoscopy devices, etc. The antenna can also be used for off-body communication to steer the radiation to the base station. The helical antenna can operate in two modes: normal mode and axial mode. In the normal mode, helix dimensions are much smaller than the wavelength ($D \ll \lambda$). Here, D represents the dimension of the helix, and λ is the operating wavelength. The antenna radiates perpendicular to the helix axis, i.e., the broadside radiation pattern is formed. The linear polarization is achieved in the normal mode operation. The helical antenna in normal mode is mostly used in applications such as implantable and RFID. In the axial mode, the dimensions of the helix are comparable to the wavelength ($D \approx \lambda$). The antenna radiates along the axis of the helix; therefore, an end-fire radiation is achieved. The antenna in axial mode normally produces circular polarization. Both RHCP and LHCP are achievable in the axial mode of operation. The broader application of helical antennas in axial mode is satellite communication and deep-space networks. The achievable gain in axial mode is 10–15 dBi. And 10%-30% fractional bandwidth is achieved in broadband operation. The input impedance of a helical antenna in free space is around 140 Ω. A quarter-wave transformer or balun is used as a matching network. One of the main advantages of using the helical antenna is the change in size. Conventionally, the helical antenna has a large size for the low frequencies, but the antenna can be made compact for miniaturized designs. The beamwidth for the axial mode operation is between 30° and 60°. If the antenna is designed properly, a high efficiency of more than 80% can be achieved. Some of the main characteristics and parameters of the conventional helical antenna are mentioned in Table 4.3

4.2.1.1 Size Considerations in Helical Antenna Design

Similar to other antennas, the size of a helical antenna depends on the wavelength (λ) of operation, i.e., on the operating frequency (f). In axial mode operation, if the diameter is represented by D and S represents the pitch of the helical antenna, then both of the parameters must be proportional to the wavelength as given in eqs. (4.1) and (4.2):

$$D \approx \frac{\lambda}{\pi} \tag{4.1}$$

$$S \approx 0.2\lambda - 0.3\lambda \tag{4.2}$$

4.2.1.2 Gain of the Helical Antenna

Low gain is achieved while operating in normal mode (~0–3 dBi). In the case of the axial mode, a high gain of ~10–15 dBi is achievable. The gain of the helical antenna depends on the number of turns. The gain can be increased if the number of turns is increased. Equation (4.3) is used to relate the gain to a number of helical turns. And the gain increases with the number of helix turns (n) according to the eq. (4.3).

$$G = \frac{15nC^2S}{\lambda^3} \tag{4.3}$$

In the above equation, C is the circumference of the helix and can be given as $C = \pi D$. S represents the spacing between turns, and λ is the wavelength. More turns lead to higher directivity and gain, making the antenna applicable in long-range communication.

Table 4.3 The default range of parameters for a conventional helical antenna

Parameters	Typical values
Gain	10–15 dBi (axial mode)
Bandwidth	10%–30% fractional bandwidth
Input impedance	~140 Ω
Beamwidth	30°–60° (axial mode)
Efficiency	>80%

4.2.1.3 Bandwidth of the Helical Antenna

The helical antenna typically has 10%–30% fractional bandwidth. It falls in the wideband category. The bandwidth depends on the pitch angle (α) and the number of turns. The larger the pitch angle (12°–14°), the higher the bandwidth. The formula used to calculate the fractional bandwidth is given by eq. (4.4).

$$\text{Bandwidth} = \frac{f_{\text{max}} - f_{\text{min}}}{f_{\text{center}}} \tag{4.4}$$

where f_{max} and f_{min} have higher and lower cutoff frequencies.

4.2.1.4 Pitch Angle of the Helical Antenna

The pitch angle is defined as the angle between a turn of the helix and the base plane. And it is calculated by using the formula:

$$\tan(\alpha) = \frac{S}{C} \tag{4.5}$$

Here, the α represents the pitch angle. S is the spacing between turns, and C represents the circumference. The optimal pitch angle is achieved at 12°–14° for the axial mode. At this pitch angle, the best gain and radiation efficiency are achieved. If we increase the pitch angle from 14°, the bandwidth of the antenna will increase, but at the cost of lower gain.

4.2.1.5 Directivity and Beamwidth of Helical Antenna

The directivity of the antenna increases with the number of helix turns of the antenna. The half-power beamwidth (HPBW) in axial mode can be approximated using the formula:

$$HPBW \approx \frac{52}{\sqrt{ND\lambda}} \tag{4.6}$$

Here, N is the number of turns. It can be observed from the above formula that increasing the number of turns can reduce HPBW, and therefore, the antenna beam becomes narrower, i.e., higher directivity.

4.2.2 Advantages of Using the Helical Antenna in Biomedical Applications

The helical antenna offers many advantages over other conventional antennas, such as circular polarization, high efficiency in biological tissue, wide bandwidth, low-power consumption, robustness against environmental change, and deep tissue penetration with good biocompatibility. The key advantages of helical antennas are as follows.

4.2.2.1 Offers Compact Design and Miniaturization Techniques

Compared to other antennas, the helix shape of the antenna reduces its size. Further, the pitch angle, diameter, and number of turns can be varied to optimize the size of the antenna, thus making the antenna more compact. Different miniaturization techniques, such as dielectric loading, can be implemented in the helical antenna. The ceramic materials can be used for dielectric loading and to further reduce the size of the antenna. The use of ceramic materials can also increase the effective permittivity. For the MedRadio band operating at 402–405 MHz, a 5–10 mm size helical antenna can be designed.

4.2.2.2 Circular Polarization

Another key benefit of using a helical antenna is that it can provide circular polarization. Polarization helps achieve strong and stable communication, and the antenna is implanted inside the body. It can also overcome the signal distortion and fading caused by multipath. The helical antenna can produce both right-hand and left-hand circular polarization, which can improve link reliability. This is the reason why a helical antenna is suitable for the implant-to-off-body device communication.

4.2.2.3 High Efficiency as an Implantable Antenna

Compared to conventional antennas such as patch antennas and dipole antennas, the helical antenna offers very good efficiency if it is implanted inside human tissue. The helical antenna offers a longer electrical length and fewer losses, reducing surface waves. This makes the helical antenna suitable for operating in low-power devices. The low-power operation is critical for battery-operated implantable devices.

4.2.2.4 Suitable for Wideband Operation

The helical antenna offers a wide bandwidth that supports a high data rate compared to conventional implant antennas. The helical antenna is suitable for operating in the MedRadio band, ISM band, and WMTS band. Therefore, it is suitable for implantable medical telemetry, wireless health monitoring, and hospital-based monitoring of the patient. The wide bandwidth of the antenna not only offers a high data rate but also reduces latency. Therefore, the antenna becomes suitable for real-time monitoring such as ECG, EEG, and glucose monitoring.

4.2.2.5 Biocompatibility and Deep Tissue Penetration

The helical antenna offers excellent biocompatibility, which is helpful for deep tissue penetration. The helical antenna offers excellent biocompatibility, which is helpful in deep tissue penetration. The helical antenna has better penetration through human tissue at lower frequencies (MedRadio: 402–405 MHz, WMTS: 1395–1400 MHz) compared to high-frequency antennas like mm-wave. At such low frequencies, the wavelength is longer, which ensures low absorption and, therefore, less heating of the tissues. Further, to improve the safety of patients, implanted devices are encapsulated with biocompatible materials such as PEEK, PDMS, Parylene-C, and ceramic coatings.

4.2.2.6 Patient Safety and Low SAR

The helical antenna can achieve low SAR as it operates at low frequencies. The radiation absorption is less at lower frequencies. Lower absorption means lower heating of tissue, and therefore, more radiation can come out of the patient's body. The SAR is a measure of the level of radiation absorption per unit of tissue. Further, low-loss dielectric materials can also be used to minimize RF absorption. This will improve radiation efficiency [9, 10].

4.2.2.7 Consumes Less Power and Is Thus Suitable for Battery-Operated Devices

The helical antenna offers high efficiency. The power losses in tissues are less. This extends the battery life of implantable medical devices. The need for frequent replacement of the batteries in implant devices can be avoided. Sometimes, the wireless power transfer (WPT) mechanism is also used to energize the implanted devices wirelessly. The inductive and capacitive WPT mechanisms are popular and discussed in detail in Chap. 5.

4.2.2.8 Robustness Against the Biological Environment Changes

The helical antenna performance does not vary with the change in the biological environment, such as muscle, blood, bones, ear canal, cerebrospinal, and thick and thin tissues. Various characteristics of the helical antenna remain intact in these varying fluid environments [11].

4.2.3 WBAN applications

- Patient monitoring: In recent years, telemedicine has been a prominent application of health care communication. The sensors to measure heart rate, temperature, ECG, EEG, oxygen level, and glucose level are placed near the patient, and then, by the use of cloud-based systems, these are transmitted to doctors for remote diagnosis.
- The application of WBAN is not limited to diagnosis but can also be used in chronic disease monitoring, such as hypertension, diabetes, or Parkinson's disease. Continuous monitoring of blood glucose, blood pressure, and body movements is done for these diseases.
- Use in pacemakers and other implanted devices: The WBAN application also includes wireless pacemakers, neurostimulators, and defibrillators. The WBAN is implemented to communicate with the off-body base station. The remote tracking of other types of implanted devices is possible with the WBAN.
- Enhance the smooth exchange of information between wearable or implantable devices and external systems. That ultimately reduces the regular hospital visits to the doctor.
- Facilitate strong data transfer for immediate health surveillance, particularly in smart hospitals for real-time patient tracking. This helps in sending a quick alert to doctors, and sometimes, the response in such scenarios becomes lifesaving.
- Another less focused application of WBAN is in sports and fitness. With the advancement of competition and modern trends, it is necessary to track the performance of athletes, which helps in the prevention of injuries and improves the experience of personalized fitness coaching. For this purpose, wearable fitness trackers such as smart watches, Garmin, and Fitbit use WBAN sensors. In sports activities, the athletes are also tracked for sleep patterns, heart and oxygen levels, and stress levels. This opens a potential field of research in AI-driven fitness monitoring. This ultimately improves the sports performance.
- Another benefit of using the WBAN in sports applications is that it can help to detect injuries. Ligament injury and muscle fatigue are common problems faced by athletes during sports activities. Physical therapy and rehabilitation can be assisted with the WBAN. The recovery of a sports person can be monitored by using WBAN.
- Applications of WBAN are also explored in the military and defence. The WBAN can give signals of soldiers' health monitoring, such as heart rate, blood pressure

level, hydration level, heat stroke, mental fatigue, etc. Injury detection and battlefield communication are other useful applications in defence.

- The abovementioned uses are one aspect of applications in the military and defence. The other aspect is smart military uniforms and other war wearables. The use of sensors, GPS trackers, smart helmets, and heating jackets is very useful in war-like situations.
- The next prominent application of WBAN is smart homes where the elderly and disabled live independently. The automated health monitoring is set up there. The WBAN devices are used to detect the fall of persons and give voice-activated alerts to the person under surveillance. For the same persons, AI-activated appliances such as light air conditioners can also be involved in the network.
- Another major application of WBAN is in disaster and emergency response. The rescue teams and firefighters can monitor life signals from the WBAN devices. Technology has become very attractive in disaster-prone geographies. For example, Japan is regularly affected by mid and high-magnitude earthquakes. The implementation of WBAN becomes very helpful in such scenarios.
- The application is not restricted to the geographies but is also helpful to the persons involved in the rescue operations. Their safety can be monitored and ensured by implementing WBAN around them. The continuous checking of the oxygen level of firefighters is an important task and can help in fire-like disasters.
- The applications of WBAN are not limited to the abovementioned scenarios. There are many other small applications useful across different fields, such as capsule endoscopy.
- Wearable devices for tracking animals in the deep jungle and underwater to maintain their count and safety. The WBAN is created through the implanted devices. The implanted devices can also have different categories. One such category, ingestible medical devices, is worth mentioning here. These are capsule-like devices, which are swallowed from the mouth like a regular pill. These devices collect images and other information while travelling through the digestive system. The collected data is regularly sent to the off-body base station. Once its work is done, it can come out in the form of feces from the human body. The WBAN network is helpful in setting up a network between the ingestible and the off-body station. The ingestible capsules are useful in medical imaging. In such a WBAN network, a data rate up to 2 Mbps is achievable [11–13].

4.2.4 Challenges in WBAN Applications

- In WBAN, the proximity of the human body to the devices produces many challenges. The WBAN devices need to communicate with an off-body control system, and hence, an antenna is an integral part of the WBAN device. As the antenna radiates in proximity to the human body, some of the radiation can reach the human body. Even in extremely directive antennas, radiation leakage because of the existence of minor and back lobes of the pattern gets absorbed by the human body and causes tissue heating and several other serious health implications. Special measures have to be taken to reduce tissue heating. The measure to observe the tissue heating is given by the specific absorption ratio (SAR). That is measured by the amount of heating per unit of tissue volume. Therefore, SAR quantifies the absorbed power by the human tissue. A threshold of SAR is set as 1.6 W/kg averaged over 1 gram of tissue. Therefore, while designing wearable antennas, implant antennas, ingestible antennas, and antennas for capsule endoscopy, the SAR limit plays a pivotal role [5].

- Another challenge in the WBAN devices is that the patient's motion or different postures can affect the performance of the antenna. Consider the case of the wearable antenna, bending, stretching, or twisting the cloth can result in distorting the antenna geometry. The antenna's physical dimensions can also be altered. And more importantly, radiation patterns can be blocked or distorted. Since the WBAN has a limited coverage range, radiation blocking or distorting can adversely affect the results obtained. Sometimes, the body's postures and movements break the antenna if the material used in the design is not flexible or sturdy enough. The challenges are not limited to wearable antennas only; they can also appear in implant antennas, where the positioning and radiation direction of the antenna should be very precise to get a fruitful result. The details of the implant antenna challenges are discussed in detail in Chap. 5. Therefore, even using the same material for antenna fabrication, the off-body antenna performance is much better compared to WBAN antennas.
- The next major challenge with the WBAN devices is the effect of the environment and surroundings in which the device is placed. The WBAN antenna's performance is affected by the temperature of the surroundings, resistance to physical abrasion, and the operational stress caused by the environment. The environmental conditions affect the functioning of the implanted devices. The type of tissue, the thickness of tissues, the density, and the ion concentration of the liquid in the tissue can affect the performance. To overcome the issue, one solution is to encapsulate the device before implanting it inside the body. An antenna array can be used in high-density or thick tissues so that the radiation can come out of the tissue. To overcome the problem of tissue and bone thickness, beam steering or pattern reconfigurable antennas can also be used. Since the communication range of implanted devices in WBAN is limited to a few meters only, the placement of the antenna should be very precise, along with its orientation. So, the maximum radiation should come out of the human body. A sufficient data rate should be achieved.

The operating frequencies of WBAN devices can raise another key challenge. Most WBAN devices operate in the ISM band as they are available free of cost. It can cause co-channel interference as many WBAN devices operate in the same band. To overcome the issue, a frequency reconfigurable antenna can be used. The design approaches and other specifications of the frequency reconfigurable antennas are discussed in Chap. 9.

4.3 Types of Helical Antennas for Biomedical Applications

The helical antenna is preferred in biomedical applications as it fits on-body and off-body communicating devices, biosensing, capsule endoscopy, etc. Their compact size and structure made them suitable for such applications. In the design of a helical antenna, the conducting wire is wound into a helical shape. To design the helical antenna, three key elements must be considered: a helix wire that is a metallic conductor wound into a spiral shape, a flat metal surface ground plane to support the antenna and reflect the radiation, and finally, a feeding line through which the power is supplied. Normally, a microstrip line or a coaxial cable is used for this purpose [14].

Helical Antenna Geometry: The key geometrical parameters for the helical antenna design are the diameter of the helix represented by D, the diameter of the conducting wire d, the pitch of the antenna S (distance between two consecutive helix turns), and the total number of turns n. The geometry of a conventional helical antenna is shown in Fig. 4.3.

The total length and axial length of the wire used in the design of a helical antenna can be calculated by using eqs. (4.7) and (4.8).

$$\text{Total length of wire} = nL \tag{4.7}$$

$$\text{Total axial length}\left(A\right) = nS \tag{4.8}$$

Here, the length L of the wire can be calculated by making a rectangular triangle flared by an angle α and mentioned in eq. (4.9). Figure 4.4 is used in the calculation of L.

$$L = \sqrt{(S^2 + C^2)} \tag{4.9}$$

From the above figure, the flair angle can also be calculated as given by eq. 4.10.

$$\alpha = \tan^{-1}\left(\frac{S}{\pi D}\right) = \tan^{-1}\left(\frac{S}{C}\right) \tag{4.10}$$

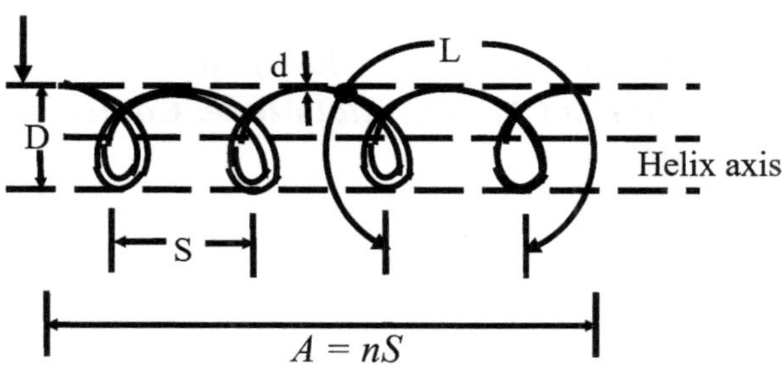

Fig. 4.3 The helical antenna design

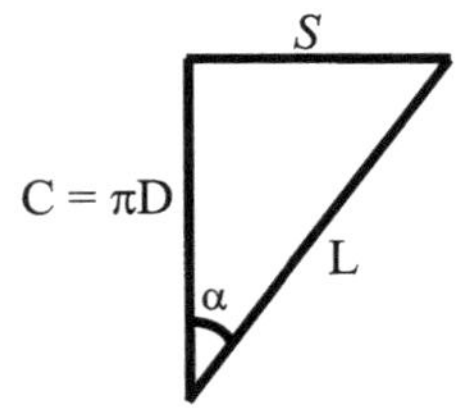

Fig. 4.4 Equivalent right-angle triangle

Table 4.4 Broader classification of helical antennas

Categorization parameters	Type of helical antenna
Based on the mode of operation	• Normal mode • Axial mode • Conical mode
Based on geometry and shape	• Monofilar • Bifilar • Quadrifilar
Based on the used material	• Wire helix • Microstrip/planar helix • Flexible/conformal helix
Other	• Conical helix • Printed helix

From the abovementioned concepts, it is clear that two cases can be derived from the helical antenna. The cases are mentioned in eqs. (4.10) and (12).

$$\text{Case 1}: \alpha = 0°; S = 0: \text{Loop Antenna} \tag{4.11}$$

$$\text{Case 2}: \alpha = 90°; D = 0: \text{Linear Antenna} \tag{4.12}$$

From the classification point of view, helical antennas are segregated among the types, such as mode of operation-based, structure-based, and the type of material used in the design of the helical antenna. Each type of helical antenna has its specific applications, operating frequency, and advantages. The same is mentioned in Table 4.4.

4.3.1 Modes of Operation-Based Classification of Helical Antenna: Normal Mode, Axial Mode, Conical Mode Helical Antenna

The helical antenna normally operates in three modes: normal mode, axial mode, and conical mode. The details of all three modes with their possible radiation patterns are mentioned in Table 4.5 [14].

4.3.2 *Geometry and Shape-Based Classification of Helical Antennas: Monofilar, Bifilar, and Quadrifilar Helix Designs*

4.3.2.1 Monofilar Helix

As the name suggests, mono means single. The antenna consists of a single conducting wire wound to make helices. Sometimes, a single printed trace is also used for the monofilar design. The antenna is mounted on a ground plane. The ground plane can be flat, shallow cup-shaped, or deep conical, as mentioned in Fig. 4.5.

The monofilar mainly works in two modes: normal and axial. In normal mode, since the size of the antenna is small compared to the wavelength, it acts like a short dipole antenna, and an omnidirectional radiation pattern is achieved. In the axial mode of operation, the size of the antenna is comparable to the wavelength, and a directional radiation pattern along the axis is obtained. The obtained beam is circularly polarized. The advantages we get from monofilar are simple design,

Table 4.5 Classification of helical antenna based on the operation mode

Mode	Condition	Structure	Radiation pattern	Details
Normal mode	$C = \pi D \ll \lambda$		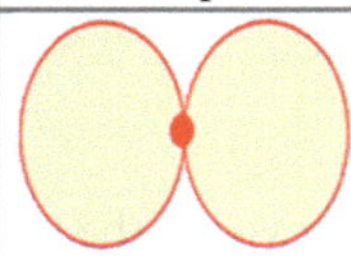	An omnidirectional radiation pattern similar to a dipole antenna is achieved
Axial mode	$C \approx \lambda$		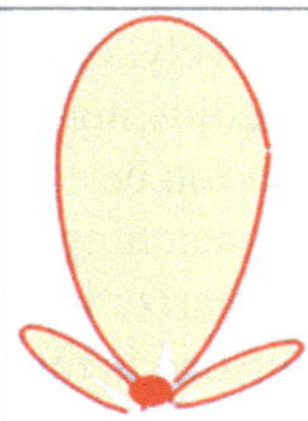	The antenna radiates along the direction of the axis of the antenna, and thus a directional pattern is achieved
Conical mode	$C \approx n\lambda$, $n = 2, 3\ldots$		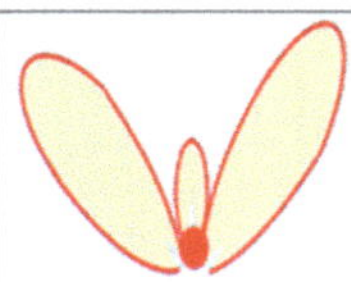	The antenna behaves in an intermediate mode with a conical beam pattern

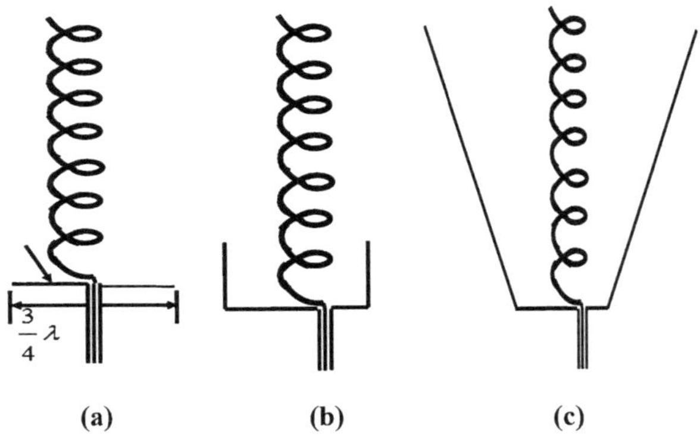

Fig. 4.5 Different ground planes in monofilar helical antenna. (**a**) Flat ground plane, (**b**) shallow cupped ground plane, (**c**) deep conical ground plane

compactness, ease of fabrication, and lightweight. The wide bandwidth can also be achieved by using the antenna. The antenna suffers from some serious drawbacks, such as lower gain compared to quadrifilar and narrow beamwidth compared to bifilar and quadrifilar [15].

4.3.2.2 Bifilar Helix

As the name indicates, Bi means two, so the two helical conductors are used in the design of the antenna. Both antennas are placed around the same axis. Since two helices are used, the current supplied to the helices can be in-phase, which means the radiation can be added constructively, or out-of-phase, which means the resultant radiation is different. Thus, by adjusting the phase of the supplied current, the radiation pattern and the polarization can be controlled. The bifilar helix is generally used to achieve better impedance matching and the axial ratio. The axial ratio decides the quality of the circular polarization. Therefore, better circular polarization is achieved in the bifilar antenna than in the monofilar. That ultimately improves the radiation efficiency. Further, the broader bandwidth can be achieved in bifilar. A caution in feeding is needed in the design of the bifilar antenna. Thus, the design is complex [16].

4.3.2.3 Quadrifilar Helix

Quadri refers to four. The antenna consists of four conductors wound in a helical shape. All the helices are placed around the axis. Because of the four helices, a phasing network is used for feeding. Each helix is fed with a supply that is typically 90° apart from the other. Thus, each helix carries a current that has a phase difference of

90° with the others. The superposition of radiating fields from all four helices can create a very pure right-hand circular polarization (RHCP) or left-hand circular polarization (LHCP). The direction of polarization is controlled by the feed. Further, both omnidirectional and directional radiation patterns are achievable. In quadrifilar, even a 0 dB axial ratio can be achieved, which means the circular polarization is extremely pure. The quadrifilar offers many advantages, such as dual resonance, wide beamwidth, better impedance matching, and compactness at the cost of a complex feeding network and larger dimensions than the monofilar and bifilar.

4.3.3 Material-Based Classification of Helical Antennas

The classification of material-based antennas is very simple. The antennas are segregated among wired helical antennas, where the metallic wire is made of copper, silver, or gold and used in traditional helix design. These antennas have applications in the biomedical and satellite communication fields. Another way to make the helical antenna is by printing it on printed circuit boards (PCBs). The designed antenna is very compact and also useful in mass production. Because of its shape, helical antennas are suitable for wearable device applications and other flexible devices. One more category is the design of flexible helical antennae on flexible substrates such as PDMS and polyimide films. The biggest advantage of the flexible substrate is that the designed antenna can be implemented on non-flat surfaces and, hence, is suitable for wrapping around the human body in wearable antenna applications. The antennas can be implanted as smart bandages or skin patches. Further, some antennas can be designed at terahertz frequencies and have applications in the biomedical field. The range of terahertz frequencies is between 0.1 and 10 THz. At terahertz frequencies, it is better to use some unconventional materials as substrate. This means that the conventional FR4, RT Duroid, etc. are not used at terahertz frequencies.

4.4 Example for Terahertz Microstrip Antenna Substrates

At terahertz frequencies, the conventional antenna faces higher losses because of the high frequencies and changes in material behavior. The surface waves become dominant at terahertz frequencies. Therefore, to neutralize these issues, the substrate choice becomes very crucial. A material with low dielectric losses (low tan δ) reduces the energy losses and provides precise control over the dimensions of the antenna and resonance. The surface of the ground plane should be smooth, as the scattering losses increase at terahertz frequencies. The material should have low dispersion to nullify the drastic changes in resonant frequency. At terahertz

Table 4.6 Properties of terahertz microstrip antenna substrates

Substrate	Electrical properties	Details
Quartz (fused silica)	Relative permittivity ($\varepsilon_r = 3.8$) Loss tangent ($\tan \delta \sim 10^{-4}$)	Offers good optical transparency. The material is highly stable and shows very low absorption. Application in THz imaging
High-resistivity silicon (HR-Si)	Relative permittivity ($\varepsilon_r = 11.9$)	Offers excellent mechanical strength to the devices
Gallium arsenide (GaAs)	Relative permittivity ($\varepsilon_r = 12.9$)	Used to design active THz components
Polyimide (flexible polymer)	Relative permittivity ($\varepsilon_r = 3.5$)	Offers flexibility. The cost of the material is low and best used to fabricate a wearable terahertz antenna. The biomedical sensors and other wearable devices are made with this substrate.
Benzocyclobutene (BCB)	Relative permittivity ($\varepsilon_r = 2.65$)	Used to design high-performance terahertz antennas

frequencies, the preferred materials are quartz (fused silica), high-resistivity (HR) silicon, Gallium Arsenide (GaAs), Polyimide (Flexible Polymer), benzocyclobutene (BCB), and Low-Temperature Co-fired Ceramics (LTCC). Out of all the above-mentioned substrates, polyimide and BCB are among the most popular to design the microstrip antenna. The polyimide is a flexible polymer material and hence suitable for wearable antenna design. The BCB is also used in high-performance antenna designs. The rest of the materials can also be useful for terahertz antenna designs. The key properties of these materials are mentioned in Table 4.6 [17, 18].

4.5 Miniaturization Techniques for In-Body Helical Antennas

The helical antenna size is bigger than that of the other conventional antennas used for implant or wearable applications. Therefore, miniaturization techniques are needed to reduce the size without affecting the performance of the antenna. Miniaturizing helical antennas for biomedical applications employs multiple techniques such as using high permittivity dielectric materials, shaping the helix, utilizing dielectric loading, modifying the ground plane, and incorporating slots or meander lines [19].

There are several miniaturization techniques as given below:

4.5.1 Shaping-Based Techniques

Helical wires can be reshaped to reduce the axial height of the helix while preserving the total wire length and operating frequency. By applying twist-and-glide symmetries in the helix design, miniaturization can be achieved without compromising radiation characteristics.

4.5.2 Dielectric Loading

Cylindrical supports made of dielectric material with a medium to high dielectric constant can enable shorter wires and a reduced overall antenna height. Although dielectric loading facilitates manufacturing, it may impact radiation efficiency due to inherent dielectric losses.

4.5.3 Ground Plane Modification

While in-body antennas are often modelled with an infinite ground plane, real-world implementations utilize finite ground planes. Reducing the ground plane size to slightly exceed the patch dimensions can aid in achieving miniaturization.

4.5.4 Other Approaches

Some other methods, such as frequency scaling (using higher frequency bands of MICS and ISM), meandered or folded helices, and metamaterials as substrate, can also be used to implement miniaturization in the helical antenna design [20, 21].

4.6 Frequency Selection for WBAN Helical Antennas

On the basis of communication link setup, the WBAN includes a variety of applications such as implant-to-implant communication, implant-to-body surface communication, implant-to-off-body communication, and body surface-to-body surface communication. The uses of antennas in these WBANs are also frequency-dependent. The different frequency ranges applicable to WBAN are given in Table 4.7 [22, 23].

Table 4.7 Various types of Wireless Body Area Networks (WBAN) and corresponding preferred frequency bands

WBAN application	Preferred frequency bands
Implant-to-implant communication	402–405 MHz (MICS band)
Implant-to-body surface communication	402–405 MHz (MICS band)
Implant-to-off-body communication	402–405 MHz (MICS band)
Body surface-to-body surface communication (line of sight)	400, 600, and 900 MHz (WMTS), 2.4–2.48 GHz (ISM), 3.1–10.6 GHz (UWB)
Body surface-to-body surface communication (non-line of sight)	400, 600, and 900 MHz (WMTS), 2.4–2.48 GHz (ISM), 3.1–10.6 GHz (UWB)
Body surface-off-body communication (line of sight)	900 MHz (WMTS), 2.4–2.48 GHz (ISM), 3.1–10.6 GHz (UWB)
Body surface-off-body communication (non-line of sight)	900 MHz (WMTS), 2.4–2.48 GHz (ISM), 3.1–10.6 GHz (UWB)

4.7 Problem Formulation

The main objective of this chapter is to design a helical antenna for reliable wireless body area network communication. The antenna should fall into the ISM frequency band of 2.4–2.48 GHz. In WBAN communication, the frequency range is suitable for body-surface-to-body-surface communication (both line of sight and non-line of sight) and also for body-surface-to-off-body communication (both line of sight and non-line of sight). These applications require an efficient communication system to transmit data to off-body devices for monitoring and diagnostics. The antenna size and biocompatibility are two significant challenges that must be taken care of. Therefore, the antenna should:

- Operate within the frequency ranges suitable for **ISM bands** of 2.4–2.48 GHz.
- Ensure stable performance despite changes in the movement of the body and changes in posture.
- Achieve sufficient **bandwidth**, **efficiency**, and **low Specific Absorption Rate (SAR)** for safe use in WBAN application.

4.7.1 Literature Survey

The design of antennas for WBAN applications has been extensively explored in this section. In the literature survey, the papers are explored based on their compactness, biocompatibility, and efficient operation across the different medical communication bands. Key studies include the literature survey mentioned in Table 4.8.

Table 4.8 Literature review

Paper	Type of helical antenna	Key feature	Details
[24]	Strip-based helical antenna	3D printed	A copper strip and conductive paint-based stripped helical antenna are designed for CP across a wide axial ratio (AR) bandwidth. Impedance matching is achieved through the non-conformal placement of the strip. The antenna is constructed on a 3D-printed polylactic acid (PLA). Maximum gain of 9.40 dBi is observed at 2.05 GHz within the 3-dB AR bandwidth. The height of the antenna is $0.35\lambda_0$. Its low-profile design and wide AR bandwidth make this antenna suitable for space communication applications
[25]	Broadband helical-inspired end-fire antenna	28 GHz broadband and MIMO configuration for 5G pattern diversity	The end-fire antenna for 28 GHz broadband communications is developed with a multiple-input-multiple-output (MIMO) configuration to support pattern diversity applications. It features a streamlined geometrical design, inspired by the conventional planar helical antenna, while eliminating the need for vias. Fabricated on both sides of an ultra-thin high-frequency substrate (Rogers RO4003, $\varepsilon_r = 3.38$, thickness: 0.203 mm), the antenna's performance was validated through experimental measurements. Remarkably, it exhibits a broad measured impedance bandwidth of −10 dB, spanning from 26.25 GHz to 30.14 GHz, effectively covering the 5G communication frequency band and achieving a peak gain of 5.83 dB
[26]	Design, fabrication, and testing Of a helical antenna	3D printing technology	The circularly polarized, 3D-printed helical antenna operates at 5 GHz. The antenna structure is fabricated using nickel electroless plating, followed by copper electroplating. The maximum gain of the antenna is measured at 13.9 dBic in the broadside direction (RHCP), with a measured efficiency of 89.1%. These results demonstrate the feasibility of rapidly fabricating lightweight, highly efficient antennas using 3D printing technology

(continued)

Table 4.8 (continued)

Paper	Type of helical antenna	Key feature	Details
[27]	Wideband strip-helical Antenna with 1.1 turns	Wideband helical antenna with 1.1 turns	A 1.1-turn wideband strip-helical antenna was numerically and experimentally analyzed, showcasing its capability to generate a wideband circularly polarized (CP) wave with 50-ohm impedance matching. The antenna is printed on a hollow cylindrical substrate with a relative permittivity of $\varepsilon_r = 2.2$ and a thickness of h = 0.5 mm. It is directly excited using a 50 Ω coaxial cable, eliminating the need for additional impedance matching components. A ground plane positioned beneath the antenna ensures a directional radiation pattern. To validate this design, a prototype of the 1.1-turn strip-helical antenna was tested, achieving an overlapped bandwidth of 46% with a height of $0.52\lambda_0$, where λ_0 represents the wavelength in free space at the center operational frequency
[28]	Deployable helix antenna at L-band for a 1-unit CubeSat	Theoretical analysis to flight model results	The 3Cat-4 mission aims to showcase the capabilities of a 1-unit CubeSat in earth observation (EO) by incorporating a combined GNSS-R and microwave radiometer payload. A key challenge lies in designing an antenna that adheres to the 1-unit CubeSat dimensions while functioning across multiple frequency bands, including global positioning system (GPS) L1 and Galileo E1 (1575 MHz), GPS L2 (1227 MHz), and the microwave radiometry band (1400–1427 MHz). Additionally, the antenna must achieve a directivity between 8 and 12 dB, depending on the band, and maintain a front-to-back lobe ratio of at least 10 dB in L1 and L2 GPS bands. Following a trade-off analysis of possible antenna types, a helix antenna was determined to be the most viable option, as it can be compactly stowed for launch and deployed in orbit. This article details the antenna design through theoretical analysis, numerical simulations, and measurements conducted on the engineering model, and the final design and performance assessment of the flight model.

(continued)

Table 4.8 (continued)

Paper	Type of helical antenna	Key feature	Details
[29]	Axial mode triangular helical antenna	Helical antenna operating in the axial Mode	The radiation pattern of a triangular helical antenna operating in axial mode is derived based on an attenuated travelling wave along the conductor, expressed in terms of the radiation vector. Experimental measurements of the radiation patterns were conducted at various frequencies for an equilateral triangular helix, with a turn length of 50 centimeters and a pitch angle of 14°, within the frequency range of 450–750 MHz. Theoretical calculations of the radiation pattern at the mid-frequency demonstrated agreement with the experimentally observed data. Additionally, the impedance characteristics of this antenna were experimentally evaluated across the same frequency band. Both the radiation patterns and impedance characteristics of the triangular helix exhibit broadband behavior similar to that of a circular helix
[30]	Nonuniform helical antennas	High-gain achieved using quad array	A high-gain quad array of nonuniform helical antennas was designed. The design is obtained by optimizing a 3D numerical model of four nonuniform helical antennas placed above a ground plane, including a model of a feeding network, utilizing the method of moments with higher-order basis functions. The gain of one optimal nonuniform helical antenna is about 2.5 dB higher than the gain of a uniform helical antenna of the same axial length. Creating a 2 × 2 array further increases the gain up to about 6 dB. The resulting quad array fits into a box whose dimensions are 2.5 × 3.3 × 3.3 wavelengths, and the gain in the main radiating direction is about 20.5 dBi in the frequency range from 0.9 GHz to 1.1 GHz. The design is verified by measurements of a prototype of the quad array
[31]	Helical antenna for wireless	Helical antenna in the form of a screw thread	A helical antenna is an electromagnetic radiator comprising a conductive wire wound in the form of a screw thread. This type of antenna operates in two principal modes: Normal mode and axial mode. In the axial mode, radiation is maximized along the helix axis and exhibits nearly circular polarization. The design of an axial mode helical antenna operating at a resonant frequency of 2.4 GHz. It features circular polarization, making it well-suited for wireless applications even in environments with obstructions. The proposed antenna demonstrates a directional radiation pattern with a gain of 10.16 dB. Further investigation is planned to adapt this design for potential applications in underwater environments

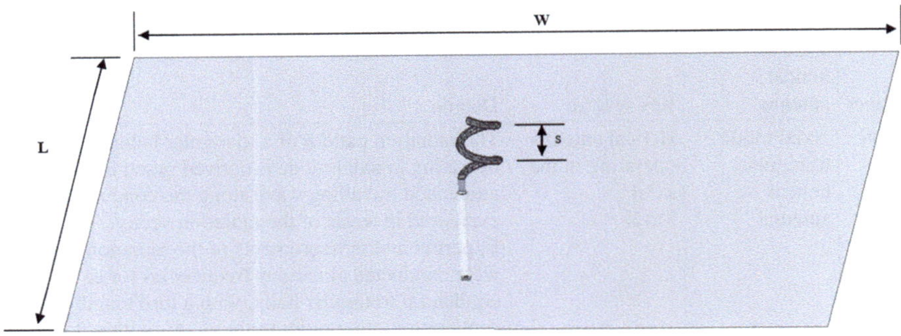

Fig. 4.6 Proposed helical antenna

4.7.2 Proposed Helical Antenna for WBAN Application

Helical antennas, developed over 70 years ago, are widely recognized for their effectiveness in the axial radiating mode, offering an almost circularly polarized wave across a broad frequency band while maintaining a relatively straightforward structure. The antenna covers a size of 12.49 × 12.49 mm^2. The helical antenna covers the 1.85 turns with helix spacing (s) of 0.62, and power is fed with coaxial. Figure 4.6 shows the proposed helical antenna design.

4.7.3 Result and Discussion

The antenna resonates at 2.48 GHz and covers the bandwidth of 2.41–2.59 GHz. The resonance and bandwidth of the antenna can be verified in Fig. 4.7, which shows the curve of parameter S11. The antenna provides an omnidirectional radiation pattern as shown in Fig. 4.8; the radiation pattern of the antenna can be justified with the 3D radiation pattern mentioned in Fig. 4.9. The resonant frequency and radiation pattern show that the antenna is suitable for wireless body area networks (WBAN) applications and will operate in the ISM band.

4.8 Helical Antenna Arrays for Enhanced Performance

The size of the helical antenna is large compared with other conventional antennas used in implant and wearable antenna applications. Therefore, different miniaturization techniques are followed. One way to reduce the dimension of the helical antenna is by implementing 2D arrays of helical antennas in place of using a long single helix antenna. The array of helices offers many advantages, such as higher gain, provides beam-steering capability, and maintains polarization purity. The

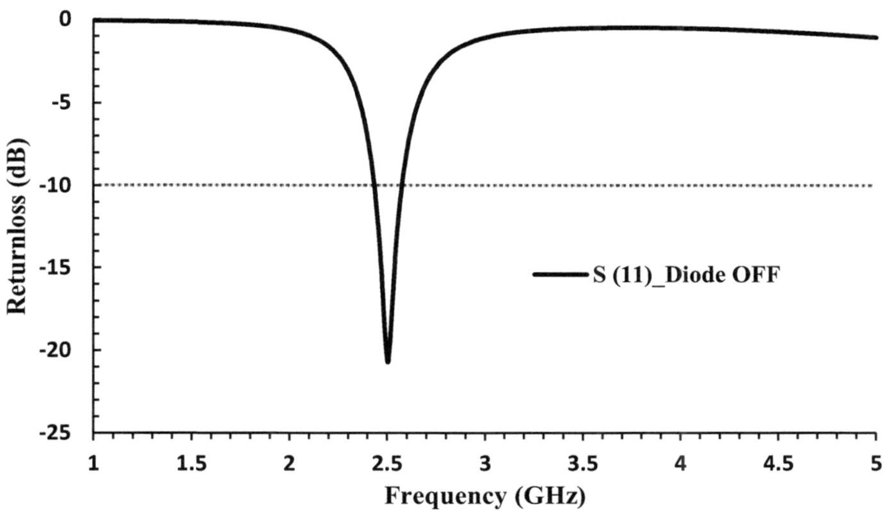

Fig. 4.7 Return loss curve of proposed antenna

helical array can be categorized as a linear array (helices are placed in a line), a planar array (helices are placed on a 2D grid), a circular helical array (helices are placed on a circular grid), and a conformal helical array (helices are placed on a curved surface). Take an example of using a 2 × 2 (four elements) planner helical antenna array in place of a single helical antenna. Let the single helix antenna have a total of 80 turns, then each helical element of the array will have 20 turns. As directivity is proportional to the number of turns, reducing the number of turns by four decreases the directivity by the same factor. Consequently, the directivity of each 20-turn helical antenna will be divided by four. To compensate for this, these four helical antennas need to be arranged in a 2 × 2 array as shown in Fig. 4.10. The figure shows the top view of the antenna array [32].

It is important to determine the optimal spacing for these helical antennas. The researchers must determine the optimal spacing among the antennas. The horizontal and vertical spacing is specified by dx and dy in the figure. To determine the efficient spacing between the antennas, first we need to calculate the aperture from the eq. (4.13) as below.

$$\text{directivity} = 4\pi \times \text{Aperture} / \lambda^2 \tag{4.13}$$

Assuming the desired value of directivity is D, substituting this value yields an effective aperture of approximately $5\lambda^2$. If we consider the aperture to be square, the side length can be approximated as $5\lambda^2$. Now consider the desired directivity D = 24 dB. Therefore, D = 24 dB = 251.19. Further, assume that the antenna is operating in axial mode. As mentioned in Sect. 4.2, the range of C_λ is between 0.8 and 1.2, and α is between 12° and 14°, so $C_\lambda = 1.05$ and $\alpha = 12.7°$ can be assumed for further calculations. Thus, using eq. (4.10), we can get

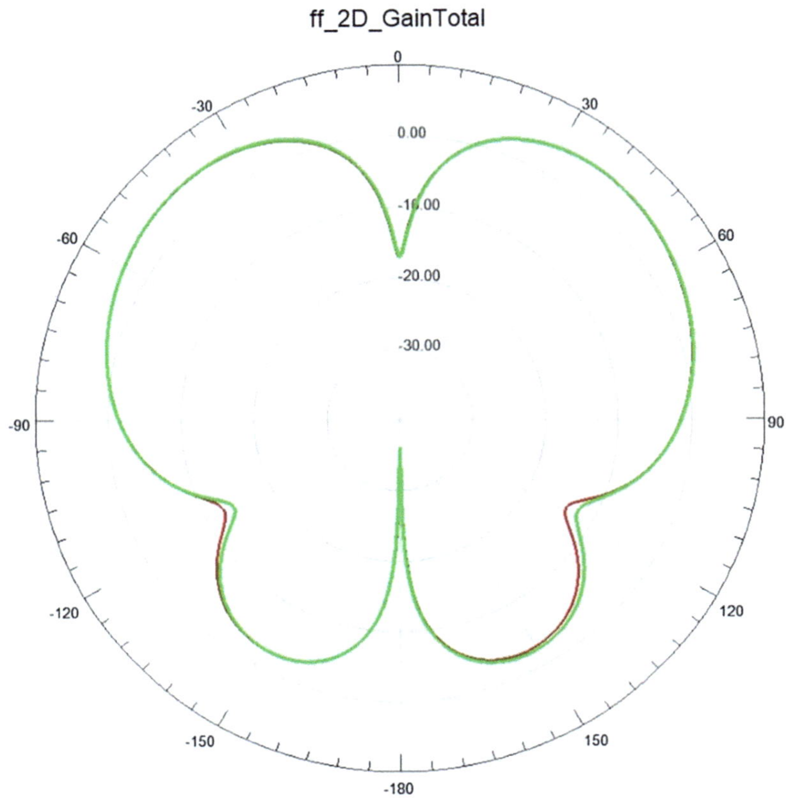

Fig. 4.8 2D radiation pattern of proposed antenna

$$S_\lambda = C_\lambda \tan\alpha = 0.2366 \tag{4.14}$$

Thus, the directivity (D) of the helical antenna can be calculated by using eq. (4.15) below.

$$D = 12 C_\lambda^2 \, n \, S_\lambda \tag{4.15}$$

After substituting the desired value of D, C_λ, S_λ in eq. (4.15), the obtained value for the number of turns (n) is given below.

$$n = \frac{251.19}{12(0.2366)(1.05)^2} = 80 \tag{4.16}$$

Now consider the case of using four 20-turn helical antennas, in place of a single 80-turn helical antenna. The directivity of each 20-turn helical antenna will

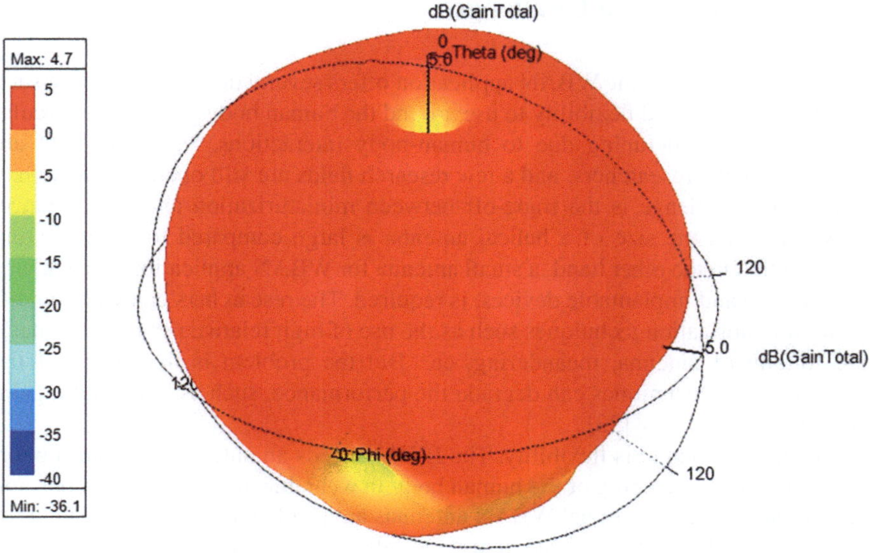

Fig. 4.9 3D radiation pattern of proposed antenna

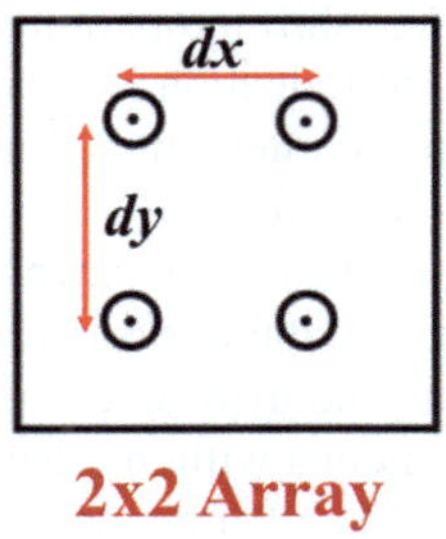

Fig. 4.10 Top view of 2 × 2 helical antenna array [32]

be = $D/4$ = 251.19/4 = 62.8. Now, using eq. (4.13), we can calculate the effective aperture.

$$\text{Aperture} = D\frac{\lambda^2}{4\pi} \approx 5\lambda^2 \tag{4.17}$$

So, if we assume the antenna array has a square aperture, it means dx = dy in Fig. 4.10. Then the effective spacing between each helix can be obtained as in eq. (4.18).

$$\text{Effective spacing} = \sqrt{5}\lambda = 2.236\lambda \tag{4.18}$$

The notable point here is that each helix is placed at the center of its aperture.

4.9 Open Research Challenges

The helical antenna in the WBAN application offers several design challenges, such as miniaturization and flexibility to use around the human body, stability of circular polarization, and detuning due to human-body interactions. Many of them are addressed by the researchers, and a few research fields are still open for research.

The first challenge is the trade-off between miniaturization and performance. Conventionally, the size of a helical antenna is large compared to the operating wavelength. On the other hand, a small antenna for WBAN application, particularly for wearable and implantable devices, is required. The researchers can work on different miniaturization techniques such as the use of high relative permittivity material, folding of antenna, meandering, etc. But the problem is that many of the miniaturization techniques can degrade the performance, such as bandwidth, gain, and antenna efficiency [33].

The next challenge is flexibility. The BWAN needs an antenna that can support the bending and stretching of the human body in wearable antenna applications. On the contrary, the conventional WBAN antennas are rigid and don't support flexibility. So, materials like textiles can be used to fabricate the helical antenna.

Another challenge is the interference caused by the interaction of the human body. Since the WBAN antennas are placed above the skin or inside the human body, the interaction of the human body can affect the performance of the antenna. The shift in operating frequency, also called detuning, change in radiation patterns, and reduction in efficiency are the problems that arise due to the effect of human-body interaction. Encapsulation or shielding is done to protect the antenna from such effects.

The helical antenna is used to achieve circular polarization. With the body movement, the purity of circular polarization degrades. The researchers need to explore the field to determine how much body movement is permissible with a helical antenna without distorting the circular polarization.

Another challenge with the use of the helical antenna for WBAN applications is that the helical antenna offers a narrow bandwidth. A careful design is required to get wideband and or multiband operation in WBAN applications [34].

Further, a few challenges are common for all implant and wearable antennas, such as maintaining the SAR within a specified limit. The SAR compliance has to be maintained by all implants and wearable antennae to protect the human tissue. All types of antennae in implant and wearable applications require integration with other electronic devices and need long-lasting battery life. Furthermore, the antennas should be robust enough to cope with changes in the surrounding environment. As in the case of the implant device of WBAN, the body fluids will change from tissue to tissue. The performance of antennas should remain unchanged even if the surrounding environment is changing.

A few challenges are common across all antenna applications, such as low-cost mass production, security, reliability, and spoof-proofing.

4.10 Summary

The chapter focuses on two-fold research area: WBAN and the use of helical antenna in WBAN application. The chapter starts with the introduction of the WBAN. The portion includes WBAN architecture, different topologies, and protocols. The design challenges of different layers in WBAN are extensively focused on in the introduction portion. Finally, different types of in-body antennas used to make a WBAN are covered in it. After the introduction, the focus of the chapter shifts to the helical antenna design. In the second section, the characteristics of a helical antenna and its advantages, applications, and challenges are covered. The third section covers the types of helical antennas. The broad categorization has been done based on the mode of operation, geometry-based, and material-based. The size of the antenna is a key challenge in the helical antenna design. Therefore, the miniaturization techniques are covered in section four. Section four covers the frequency band used for the WBAN application. In the next section, the helical antenna for the WBAN application is designed. This section starts with problem formulation and includes subsections such as a literature survey, antenna design, and results and discussion. In the chapter, a helical antenna for the WBAN application is designed, exhibiting resonance at 2.48 GHz and covering the band between 2.41 and 2.59 GHz. The frequency band falls in the range of the ISM band and is suitable for multiple WBAN applications, such as body-surface-to-body-surface for both line of sight and non-line of sight communication, and body-surface-to-off-body-communication for line of sight and non-line of sight communication. The next section focuses on antenna design validation techniques. After this, the open research challenges presented in the stated work are mentioned in the chapter.

References

1. D.M.G. Preethichandra, L. Piyathilaka, U. Izhar, R. Samarasinghe, L.C. De Silva, Wireless body area networks and their applications—A review. IEEE Access **11**, 9202–9220 (2023). https://doi.org/10.1109/ACCESS.2023.3239008
2. M. Yaghoubi, K. Ahmed, Y. Miao, Wireless body area network (WBAN): A survey on architecture, technologies, energy consumption, and security challenges. J. Sens. Actuator Netw. **11**, 67 (2022). https://doi.org/10.3390/jsan11040067
3. D. Xue, B. Garner and Y. Li, Electrically-small folded cylindrical helix antenna for Wireless Body Area Networks. 2016 Texas Symposium on Wireless and Microwave Circuits and Systems (WMCS), Waco, TX, USA, 2016, pp. 1–4, doi: https://doi.org/10.1109/WMCaS.2016.7577479.
4. I. Elfergani, A. Iqbal, C. Zebiri, A. Basir, J. Rodriguez, M. Sajedin, A.D.O. Pereira, W. Mshwat, R. Abd-Alhameed, S. Ullah, Low-profile and closely spaced four-element MIMO antenna for wireless body area networks. Electronics **9**(2), 258. https://doi.org/10.3390/electronics9020258
5. U. Ali, S. Ullah, B. Kamal, L. Matekovits, A. Altaf, Design, analysis and applications of wearable antennas: A review. IEEE Access **11**, 14458–14486 (2023). https://doi.org/10.1109/ACCESS.2023.3243292

6. A. Hegde and D. Prasad, Evaluation of PHY and MAC layer in WBAN using Castalia: Lower layer parameters of WBAN in Castalia. 2017 International Conference on Intelligent Sustainable Systems (ICISS), Palladam, India, 2017, pp. 395–399, doi: https://doi.org/10.1109/ISS1.2017.8389438.
7. A. Kiourti, K.S. Nikita, A review of in-body biotelemetry devices: Implantables, Ingestibles, and Injectables. IEEE Transactions on Biomedical Engineering **64**(7), 1422–1430 (2017). https://doi.org/10.1109/TBME.2017.2668612
8. A.B. Dey, W. Arif, On-body low-profile compact AMC-integrated wideband antenna for body area network applications. IETE Technical Review **41**(2), 133–146 (2023). https://doi.org/10.1080/02564602.2023.2220927
9. S. Vladimirov, V Vishnevsky, A. Larionov, and R. Kirichek, The model of WBAN data acquisition network based on UFP. Distributed Computer and Communication Networks: 23rd International Conference, DCCN 2020, Moscow, Russia, September 14–18, 2020, doi: https://doi.org/10.1007/978-3-030-66471-8_18.
10. L. Anil, R. Cherian, K. Saravanan, Investigation and survey of a wearable multiband antenna for ISM band applications. 2021 International Conference on Disruptive Technologies for Multi-Disciplinary Research and Applications (CENTCON), 285-288, (2021), https://doi.org/10.1109/CENTCON52345.2021.9687995.
11. S.M. Demir, F. Al-Turjman, A. Muhtaroğlu, Energy scavenging methods for WBAN applications: A review. IEEE Sens J **18**(16), 6477–6488 (2018). https://doi.org/10.1109/JSEN.2018.2851187
12. M.K. Goenka et al., Capsule endoscopy: Present status and future expectation. World J Gastroenterol. **20**(29), 10024–10037 (2014). https://doi.org/10.3748/wjg.v20.i29.10024
13. G. L. P. Ashok, G. V. Nath, B. C. Neelapu, UWB Antenna for Microwave Imaging and Communication Applications. 2024 Second International Conference on Microwave, Antenna and Communication (MAC), 1-5, (2024), https://doi.org/10.1109/MAC61551.2024.10837230.
14. J.D. Kraus, *Antennas and Wave Propagation* (Tata McGraw Hill Education, 2010) ISBN 1283186926
15. S. M. Farghaly Sayed and O. Y. K. Alani, A dual-mode helical antenna design for on-and off-body communication in body sensor networks. 2016 3rd International Conference on Soft Computing & Machine Intelligence (ISCMI), Dubai, United Arab Emirates, 2016, pp. 239–243, doi: https://doi.org/10.1109/ISCMI.2016.18.
16. R. H. Binti Mohd Baharin, Y. Yamada, K. B. Kamardin and N. Michishita, Changes of electric current on a normal-mode helical antenna inside a human body. 2017 IEEE Conference on Antenna Measurements & Applications (CAMA), Tsukuba, Japan, 2017, pp. 403-406, doi: https://doi.org/10.1109/CAMA.2017.8273465.
17. Z.R.M. Hajiyat, A. Ismail, A. Sali, M.N. Hamidon, Design and analysis of helical antenna for short-range ultra-high-speed THz wireless applications. Optik **243**, 167232., ISSN 0030-4026 (2021). https://doi.org/10.1016/j.ijleo.2021.167232
18. S. Younes, F. Jaouad, Enhancement of a THz Patch antenna performance using metamaterials for biomedical applications. In Mejdoub, Y., Elamri, A. (eds) Proceeding of the International Conference on Connected Objects and Artificial Intelligence (COCIA2024). COCIA 2024. Lecture Notes in Networks and Systems, vol 1123. Springer, Cham. https://doi.org/10.1007/978-3-031-70411-6_4.
19. T.T. Le, T.-Y. Yun, Miniaturization of a dual-band wearable antenna for WBAN applications. IEEE Antennas Wireless Propag Lett **19**(8), 1452–1456 (2020). https://doi.org/10.1109/LAWP.2020.3005658
20. T. Islam, S. Roy, Low-profile meander line multiband antenna for wireless body area network (WBAN) applications with SAR analysis. Electronics **12**(6), 1416 (2023). https://doi.org/10.3390/electronics12061416
21. T. Saeidi, M. Gokdemir, S. Karamzadeh, Metamaterial-based circularly polarized wearable antenna for ISM and 5G communications. 2022 30th Signal Processing and Communications Applications Conference (SIU), 1-4, (2022), https://doi.org/10.1109/SIU55565.2022.9864810.

22. D. Xue, B. A. Garner and Y. Li, On-body radiation of 3D-printed fold cylindrical helix (FCH) wearable antenna. 2017 Texas Symposium on Wireless and Microwave Circuits and Systems (WMCS), Waco, TX, USA, 2017, pp. 1–4, doi: https://doi.org/10.1109/WMCaS.2017.8070687.
23. R. H. M. Baharin, T. Uno, T. Arima and Y. Yamada, Input resistance and radiation efficiency of normal-mode helical antenna inside human body. 2019 International Symposium on Antennas and Propagation (ISAP), Xi'an, China, 2019, pp. 1–3.
24. P. Ghosh, F. Harackiewicz, 3D printed low profile strip-based helical antenna. Prog Electromagn Res C **127**, 195–205 (2022)
25. H. Zahra, W.A. Awan, W.A.E. Ali, N. Hussain, S.M. Abbas, S. Mukhopadhyay, A 28 GHz broadband helical inspired end-fire antenna and its MIMO configuration for 5G pattern diversity applications. Electronics **10**(4), 405 (2021)
26. B. Ghassemiparvin, N. Ghalichechian, Design, fabrication, and testing of a helical antenna using 3D printing technology. Microw Opt Technol Lett. **62**, 1577–1580 (2020)
27. X. Tang, B. Feng, Y. Long, The analysis of a wideband strip-helical antenna with 1.1 turns. Int J Antennas Propag **2016**, 5950472., 7 pages (2016)
28. L. Fernandez, M. Sobrino, J.A. Ruiz-de-Azua, A. Calveras, A. Camps, Design of a deployable helix antenna at L-band for a 1-unit CubeSat: From theoretical analysis to flight model results. Sensors **22**(10), 3633 (2022)
29. S. Balaram, T.S. Venkataraman, Axial mode triangular helical antenna. IETE J Res **12**(11), 568–576 (1966)
30. J.L. Dinkić, D.I. Olćan, A.R. Djordjević, A.G. Zajić, High-gain quad array of nonuniform helical antennas. Int J Antennas Propag **2019**, 8421809., 12 pages (2019)
31. A. N. Jaafar et al., Analysis of Helical Antenna for Wireless Application at 2.4 GHz. 2019 IEEE Asia-Pacific Conference on Applied Electromagnetics (APACE), Melacca, Malaysia, 2019, pp. 1–5.
32. Online: https://www.cdeep.iitb.ac.in/slides/A16/EE609/EE609-L17.pdf
33. L. Kannappan, S. K. Palaniswamy, M. Kanagasabai, S. Kumar, T. R. Rao, T. Govindan, CPW-fed quad-band dual port meandered monopole automotive antenna. 2022 International Conference on Communication, Computing and Internet of Things (IC3IoT), 01-04, (2022), https://doi.org/10.1109/IC3IOT53935.2022.9767894.
34. A. Chandra, S. Aswath, N. Kishore, V Varun, Dual polarized CPW-fed multi-band antenna for 5G, ISM, and X-band applications. 2024 Control Instrumentation System Conference (CISCON), 1-6, (2024), https://doi.org/10.1109/CISCON62171.2024.10696816.

Chapter 5
Planar Inverted-F Antenna (PIFA) in Wireless Medical Devices

Contents

© The Author(s), under exclusive license to Springer Nature Switzerland AG 2026

A. Segun Ayokunle et al., *Biomedical Antenna Technology*,
https://doi.org/10.1007/978-3-031-95268-5_5

5.1 Introduction

The implant devices with wireless communication capability promise large improvements in patient care and the quality of healthcare systems. The devices used in pacemaker communication, glucose monitoring, insulin pumps, and endoscopy are a few examples that help in medical treatment. A frequency band referred to as Medical Device Radiocommunication Service (MedRadio, 402–405 MHz) has recently been allocated by the Federal Communications Commission (FCC) for medical applications [1]. Among all components essential for implant applications, the antenna plays a pivotal role in establishing robust communication links between the implant device and the off-body base station [2]. The antenna helps the implant device achieve substantial miniaturization. Designing an electrically small radiating antenna typically reduces the efficiency and bandwidth in the challenging environment of the human body. The human body causes significant challenges, such as thickness, water content, and variation of dielectric and conduction properties [3]. Recent developments have introduced several implantable antennas with diverse characteristics and target applications. The applications of antennas in healthcare are not limited to implant devices, telemetry communication, wireless body area networks (WBAN), and wearable devices; there are other fields where an antenna is used in health monitoring systems. Other frequency bands, such as Industrial Scientific and Medical (ISM) and Wireless Medical Telemetry Systems (WMTS), are also used according to the demand [4]. The on-body and off-body applications of antennas are covered through these bands. The use of higher frequencies narrows the wavelength and results in the reduction of the size of the antenna. The approach looks effective for implant and wearable antenna applications, but the performance in terms of bandwidth and data rate is compromised. For a better data rate, an antenna operating at high frequencies is desirable. Therefore, a trade-off between the data rate and the size of the antenna is made [5].

The implant antenna applications are not limited to the MICS band; the ISM frequencies can also be used in vitro and in vivo tests for intracranial pressure monitoring. The antennas in implant devices can be used in the tissues inside the body or tissues taken out from the body. Another novel application of a miniaturized antenna is in the retinal prosthesis. Bone fracture detection and monitoring, as well as capsule endoscopy, are some other major applications where a miniaturized antenna is preferred, but the antenna does not operate in the MICS band. The frequency band besides MICS, which is prominently used for in-body communication, is ISM. A lot of literature is already available on this band where the communication links are investigated and modelled. The 2.4–2.48 GHz ISM band is particularly useful in the in-body medical device. The commercial-off-the-shelf (COTS) components are available for this band [6].

In the case of implant devices, the distance between the transmitter and the receiver depends on the depth of the antenna implanted inside and the on-body or off-body base station. Path losses are also incurred in such communication links.

The path losses are caused by the permittivity, current conduction capability of the tissue, thickness, and the water molecules in the tissue and between different layers of the tissues, which may disrupt the communication between the transmitter and the receiver. Therefore, the selection of an antenna for the sustained communication link is very crucial. Antennas are essential components for medical devices nowadays. Modern health applications like implantable and wearable devices need compact wireless devices. Telemedicine, medical telemetry, and monitoring have become pivotal in modern health applications. Further, the applications are extended to medical health records. In implanted medical devices, the sensors are integrated with the devices. The limited processing capability and power requirement necessitate communication between in-body implant devices and the out-body monitoring units. Take the example of an implanted antenna for a diabetic patient, the dielectric properties of human tissue, biocompatibility, glucose condition, and impedance matching are considered when designing the antenna. In other monitoring applications, electrocardiograms (ECGs), electroencephalograms (EEGs), pH values, and body temperatures are mainly included. Breast cancer detection is a prominent diagnostic application. Different types of antennas are offered for the implant devices, such as dipole, microstrip patch, microstrip, spiral, meandered, or dielectric resonators to ensure compactness. Another important antenna used for implant applications is the Planner Inverted-F Antenna (PIFA). For an implant device, the dimensions of the antenna, radiation efficiency, specific absorption rate (SAR), and power requirement are the key characteristics. Therefore, a low-profile, highly efficient, and excellent impedance matching antenna is the desired choice for these modern applications. The PIFA is an antenna used in medical applications like implants, telemetry, wearables, etc. The PIFA antenna offers compactness with reliable performance [3]. The PIFA, as an implant antenna, should be as small as possible without affecting the surrounding tissues. The planner structure is an essential feature of PIFA as it requires relatively less space. In the design of the planner antenna, the dimensions and performance are dependent on the substrate properties. With the increase in permittivity, the performance can deteriorate despite the size becoming small [7, 8].

However, in the Internet of Medical Things (IoMT) applications, the implant devices are equipped with sensors and placed inside the human body. The antennas are also fabricated with sensors in such devices. The antennas of IoMT devices pose different challenges, such as miniaturization, flexibility, and operation at a particular frequency. Efficient design and fabrication methods are required to achieve such a level of miniaturization and flexibility. Various techniques to achieve miniaturization, such as the use of high permittivity materials, shorting pins, etc., are discussed in this chapter. Every method has its own requirements and challenges. Sometimes, the reconfigurability of antennas is also desired to tune multiple frequencies and steer the beam. The reconfigurable antenna design is discussed in detail in Chap. 10.

The PIFA can be used for any type of implant and on-body application. It can be used for the IoMT devices as well. The antenna used to operate inside the human tissue has to take care of the biocompatibility of the substrate and the specific absorption ratio (SAR) level. The biocompatibility of the substrate is defined as a feature of how much the chemical characteristics of the substrate suit the tissue environment. The high SAR level always poses a big challenge in implant devices. SAR is a measure of the rate at which the RF excitation of human tissue is converted into the heating of the sample. For the calculation of the SAR, 1- to 10-gram tissue is used, and its heating behavior is checked under the RF excitation to observe the overall effect on the human body.

5.1.1 Significance of the PIFA as an Implant Antenna

Real-time monitoring: The PIFAs are used in implant applications to monitor continuously the physiological signals and send them to the outside body. This leads to the early detection of many diseases and medical conditions. Real-time monitoring sometimes helps in prompt intervention.

Enhanced mobility: The wireless communication-enabled implant device gives the patient the liberty to move around freely. The patient is not tethered to the monitoring equipment. Therefore, the quality of patient life is improved.

Minimal invasive care: The use of PIFA in implant devices reduces the frequent visits of patients to hospitals or health officials for invasive tests. The patient has to go through a one-time surgery to place the implant. This minimizes the risk of infection.

Telemedicine and remote healthcare: The implanted antenna enables health officials to monitor the patient's condition remotely. This feature is particularly useful in suburban and rural areas lacking health infrastructure. In remote areas where regular visits by patients to the hospital are not possible, telemedicine technology can be implemented.

Smart implants: The substantial use of PIFAs is seen in smart implants such as neural stimulators, pacemakers, and insulin pumps. The smart implant provides precise control of the fluctuations occurring in the surroundings.

5.2 Advantages and Limitations of Planar Inverted-F Antenna (PIFA)

The PIFA offers multiple advantages over conventional antennas. The low SAR and omnidirectional radiation pattern make PIFA an obvious choice for medical applications. The low absorption will cause reduced tissue heating. The omnidirectional radiation pattern gives freedom to the off-body base station placement.

5.2.1 Benefits of PIFAs

- The PIFAs inherit a compact design. This reduces the size of the medical device and makes it particularly useful in the design of implants and wearable devices. For implant applications, the PIFA can operate at a frequency of around 400 MHz. The size of the conventional antenna operating at this frequency is very large compared to the PIFA.
- Many applications of wearable devices require an antenna independent of orientation. The PIFA is useful because of its omnidirectional radiation pattern. PIFA ensures sustained communication irrespective of the position and orientation of the device.
- Compared to the conventional antenna, the PIFA offers high efficiency. The feature is particularly useful in medical telemetry. The energy consumption of PIFA is low. It can be powered with a battery for a longer duration of time.
- The SAR indicates the absorption of energy by the human tissue. For the implant antennas, the SAR should be as low as possible. This ensures less harm from electromagnetic exposure. The PIFA follows these safety measures.
- Easy and seamless integration with other medical components is another requirement for medical antennas. For implant antennas, the sensors collect the information that is transmitted outside the body. Along with the sensor and antenna, some other components, like low noise amplifiers (LNAs), etc., are also fabricated in implant devices. There should be a seamless integration of all these components so that the overall form factor of the device will remain intact. The PIFA offers this integration with minimal impact on the form factor. Further, a minimum connection loss is desired in the implant devices.
- A perfect impedance match is needed to reduce the signal loss from the antenna. Perfect impedance matching reduces the standing waves, which leads to better return loss performance. The PIFA offers good impedance matching to improve overall performance.
- The planner antennas are easy to build and implement in medical devices. The integration with another component on the same surface reduces the physical damage and detachment problems. With PIFA, the advantage is easily achieved.
- Most of the time, the dielectric material used in PIFA is relatively cost-effective. This becomes economical in the mass production of the implant and wearable devices. The mass production of wearables is required in military applications.

5.2.2 Limitations

- Limited bandwidth is a major setback in the case of PIFA. This is the limitation that comes with most of the implant antennas. The less bandwidth restricts the applications and the multiband operations of the PIFA. The achievable data rate is limited, and hence, the use of PIFA in real-time telemetry applications is restricted.

- The implant antennas fall in the category of extremely small antennas. The largest dimension of the antenna is approximated by $\lambda/40$. The lossy substrate is another limitation of an antenna of such a small size. An antenna with its largest dimension "a" can be defined as electrically small if it follows the eq. (5.1).

$$ka = \frac{2\pi}{\lambda} a \leq \frac{1}{2} \tag{5.1}$$

In the above equation, "k" represents the propagation constant of the medium. The small antenna stores most of its power in the near field, and a very small portion of the power is radiated in the far field. If the radiation quality factor of the antenna is Qrad. The Qrad is inversely proportional to the bandwidth. Hence, the bandwidth of the antenna reduces as the size of the antenna becomes smaller. A trade-off has to be made between the size of the antenna and the required bandwidth.

- In the proximity of human tissue, the PIFAs are prone to detuning. It not only degrades the antenna performance but also increases the chances of false alarms.
- The precise optimization is required to design the PIFA. The medical spectrum bands are very thin, and it will be very challenging to achieve resonance of the antenna in those bands.
- Low radiation resistance is another limitation of extremely small antennas, which suffer from poor impedance matching and radiation efficiency. One way to overcome the problem is to increase the radiation level by expanding the heavy current. However, because of the size of PIFA, the use of the sachem is also limited. The addition of antenna loss can increase the bandwidth at a cost of a lower quality factor. It means that in a lossy antenna, the quality factor is directly proportional to the radiation efficiency ($Q \propto \eta$). In the case of a printed antenna, the substrate adds losses and, therefore, increases the input resistance. This leads to lower quality factors with reduced efficiency [9].

5.3 Antenna Integration in Wireless Medical Devices

The integration of the antenna with the other components is a critical aspect of medical devices. The integration of the antenna directly impacts the safe positioning, performance, and reliability of the medical device. Seamless communication is the primary requirement of these medical devices. Hence, the performance matrices are crucial [10].

5.3.1 Device Components

For any wireless medical device, several aspects must be taken care into account. A broad classification of these can be made between physical constraints and radiation properties. A proper model for communication needs to be developed as it strongly influences the antenna integration. Irrespective of the application, antenna integration is useful to achieve the final structure of the communication model. The physical constraints include the volume of the entire wireless medical device (antenna, biosensors, electronics, insulation, and batteries), whether it is used at the transmitter side or the receiver side. The cylindrical housing is used to contain all components for implant application. The biosensors and batteries are integrated with the antenna. For endoscopy applications, the capsule-type casing is preferred. And the camera, along with other antennas, should be fitted in the same capsule. The wearable application prefers a printed antenna or the shape of an antenna that can be woven or knitted. The sensors are also knitted in the cloth for real-time monitoring purposes. The off-body telemetry communication device has more liberty in terms of space. Large-sized antennas can also be used here to achieve a high data rate in 5G communication. Dense packaging is required to reduce the volume of the wireless medical communication device. The feeding point of the antenna is placed at a predefined area to reduce complexity.

Biocompatible insulation is also required to ensure efficient working for implant devices. The insulation avoids the adverse tissue reaction to the devices. Most of the tissue contains sufficient water to alter the electrical properties of the antenna.

The second aspect is the radiation properties. The operation in a specific band is desired based on the application. The operating frequency is selected precisely to cater to the application. Bands like MICS, Med Radio, ISM, WMTS, UWB, and MBAN are predominantly used in wireless medical applications [11]. The popular frequency bands with correspondingly allowed channels by the FCC are given below:

- Medical implant communication service (MICS): 402–405 MHz.
- Medical device radio communications service (MedRadio): 401–406 MHz, 413–419 MHz, 426–432 MHz, 438–444 MHz, and 451–457 MHz.
- Industrial, scientific, and medical (ISM) band: 433.05–434.79 MHz, 902–928 MHz, 2.4–2.4835 GHz, 5.725–5.850 GHz.
- Wireless and Medical Telemetry Services (WMTS) bands (608–614 MHz, 1.395–1.4 GHz, 1.427–1.432 GHz).
- Ultra-wideband (UWB): 3.1–10.6 GHz.
- Medical body area network (MBAN): 2.360–2.4 GHz.

Sometimes, a dual-band operation with one resonating frequency in the ISM band is preferred for transceiver operation in wireless medical devices. The ISM band at 2.4GHz is used to awaken the device from a sleeping state, and the other band is used for data transmission/reception purposes.

In the case of implant devices, a single feeding point antenna is preferred to minimize its complexity. The antenna gains and other radiation characteristics can be tested in a body phantom. The position of the implant in the body also has little effect on EM radiation. Further, the antenna should show robustness against the presence of other electronics and batteries. To observe the impact of tissue on EM radiation, researchers are advised to use equivalent body models [6].

From the above discussion, we can ensure that the integration of antenna in medical devices requires careful consideration of the different modules and their connectivity. The main modules in any wireless medical device are as follows:

- The **transceiver modules** ensure the use of communication between in-body/on-body and out-body devices. These devices must be compatible with different protocols, including Bluetooth, Wi-Fi, and Zigbee. A reliable data exchange with a high rate and throughput is needed in telemedicine applications. Therefore, compatibility with 5G or 6G standards is also desirable.
- The **energy module** is responsible for energizing the device. In the case of implant and wearable devices, energy efficiency becomes very crucial. It is required that the device go into a sleeping state if it is not in use. Only then will the module ensure a prolonged life for the devices.
- The **PCB design module** ensures the close coupling of different components. To make the device very compact, maximum space utilization is necessary for medical devices. This must be achieved while maintaining the desired radiation characteristics.
- The **encapsulation module** is important as it requires biocompatible materials to provide shielding inside the tissue. The performance remains intact while ensuring the proper functionality and the safety of human life.

5.4 Design Considerations for Medical Applications

The design of an antenna for any medical device is application-specific. For example, in IMD applications, a near-field radiating antenna is preferred, and in telemetry applications, a far-field radiating antenna is preferred. In on-body communication applications, the distance of the base station from the patient is crucial to defining the radiation properties of the antenna. Further, the different electrical properties of different tissues can also affect the performance of the antenna in implantable applications. Further, the thickness of the tissue also matters. The limited bandwidth also added to the challenge, particularly in achieving higher data rates, and this is the reason for exploring all the available operating frequencies. For the off-body telemetry application, the dimension of the antenna is not a constraint, but other challenges, such as co-channel interference, fading, polarization, and maximum allowed communication range, are evident. The antennas used in on-body applications also need to take care of the patient's movement, positioning, orientation, polarization, etc. Therefore, the design of an antenna for medical applications has to take care of many constraints in a limited space [8]. Some of the important aspects of the antenna design are mentioned below.

5.4.1 Frequency Band Selection

The frequency bands are crucial in medical applications. Regulatory standards such as MICS, ISM, and WMTS must be followed when designing the antenna for medical applications. Table 5.1 shows different frequency bands with their corresponding frequency ranges and applications. The MICS band can support a low data rate communication and is not suitable for medical telemetry applications. The Federal Communications Commission (FCC) allows a few additional bands to be provided for faster data rate transfer. ISM and MICS are such bands that operate in the GHz range. The frequency ranges and corresponding applications are mentioned in the table below. The WMTS band is designated by the FCC to monitor the physiological conditions of the patient. The glucose level, blood pressure, pulse rate, etc. are mainly monitored through WMTS bands.

5.4.2 Bandwidth

The bandwidth is a parameter that is set to a fundamental trade-off with the efficiency of small antennas. With the increase in bandwidth, achieving good efficiency is a challenging task. The data rate and the size of the antenna are also dependent on the achievable bandwidth. A higher bandwidth is desired to achieve a high data rate.

Table 5.1 Different frequency bands with their corresponding frequency ranges and applications

Frequency band	Frequency range	Applications
MICS (medical Implant communication services)	402–405 MHz	Wireless medical telemetry, implantable devices (e.g., pacemakers, insulin pumps), continuous glucose monitoring
MedRadio band	401–406 MHz, 413–419 MHz, 426–432 MHz, 438–444 MHz, and 451–457 MHz	Cardiac pacemakers, implantable defibrillators, neurostimulators, drug delivery systems (insulin pumps), smart prosthetics, continuous glucose monitoring, remote patient monitoring systems
ISM band	2.4–2.5 GHz	Short-range communication, wearable health monitoring devices, wireless medical imaging, and real-time health tracking
WMTS (wireless medical telemetry service)	608–614 MHz, 1395–1400 MHz, 1427–1432 MHz	Patient monitoring systems in hospitals include ECG, pulse oximetry, and respiration monitoring

In the case of an implantable antenna, the frequency of operation is in the range of 0.4 GHz. At this frequency, the achievable bandwidth becomes very limited. On the contrary, if the antenna is tuned at a higher frequency, the SAR of tissue becomes high. Sometimes, this becomes very hazardous. In power transfer applications, bandwidth is not very important. The reduction of bandwidth can be tolerated in the design stages. Therefore, if the base station is placed near the patient and the application does not demand a very high data rate, then it is good to have a sharp bandwidth to achieve greater efficiency. [8]

5.4.3 Radiation Efficiency

Radiation efficiency is a crucial parameter in the medical antenna application. The antenna efficiency is defined as the effectiveness of converting the incoming power into EM radiation. For off-body applications, achieving high radiation efficiency is relatively easy compared to implantable applications. The reason is that the surrounding tissues absorb the weak radiation. The energy will come out from the dispersive tissues to open space. The process is influenced by many frequency-dependent mechanisms, such as conductive losses, input power reflection, dielectric losses, and losses caused by the impedance contrast. There are many solutions proposed in the literature for the abovementioned issues, such as optimization of bandwidth, dielectric loading to the antenna, and the use of a high permittivity substrate. Sometimes superstates can also be used.

5.4.4 Low-Power Consumption

The antennas require sufficient power to drive the radiation up to a longer distance. In telemetry applications, a very large coverage is required. And because of off-body installation, the power requirement is satisfied by different energy sources. Power consumption has become a major constraint in implantable and wearable antennas. The source of energy supply is mostly lithium-ion batteries, and its replacement looks very difficult in implantable applications. In wearable applications, the knitting of batteries along with the antenna is very tough. The performance is limited to the power consumption. Several approaches have been followed in the literature to recharge the battery, such as an inductive loop. In the case of telemetry applications, if medical devices are not in use, they will go into sleep mode to save power. Further, transceivers are used to make power consumption optimum. The transceiver practices two signals: one signal is used in the wake-up mode of the device. In such a case, the 2.4 GHz ISM band is used for the transmission. The other signal is the low-power consumption sleep mode. The same process is followed for implantable applications. The IMD will remain asleep until it wakes up. After waking up, the device will transmit the message in the MICS or MedRadio band. Once the transmission is finished, the device will again go into sleep mode. Sometimes, these events are organized according to the doctor's schedule. An externally controlled program will make the device active for transmission.

There are many wireless power transfer (WPT) techniques available on the market that can be applied to recharge the battery of implant antennas. The WPT techniques are preferably used for pacemakers, capsule endoscopy, brain simulator, etc. The WPT techniques eliminate the power capacity barriers and challenges of surgically replacing the battery. Figure 5.1 shows the classification of the wireless power transfer methods. Based on the EM field, the WPT techniques are broadly classified into radiative far-field WPT and non-radiative near-field techniques. The recharging range is the key difference between the far-field radiative and near-field non-radiative WPT. In the case of non-radiative, the transmission distance varies from millimeters to a few meters. For radiative far-field WPT, the range can be extended up to several kilometers. Radiative far-field WPT techniques are further classified into microwave power transfer and laser power transfer. The non-radiative near-field WPT techniques are further classified based on the coupling mechanism. Inductive,

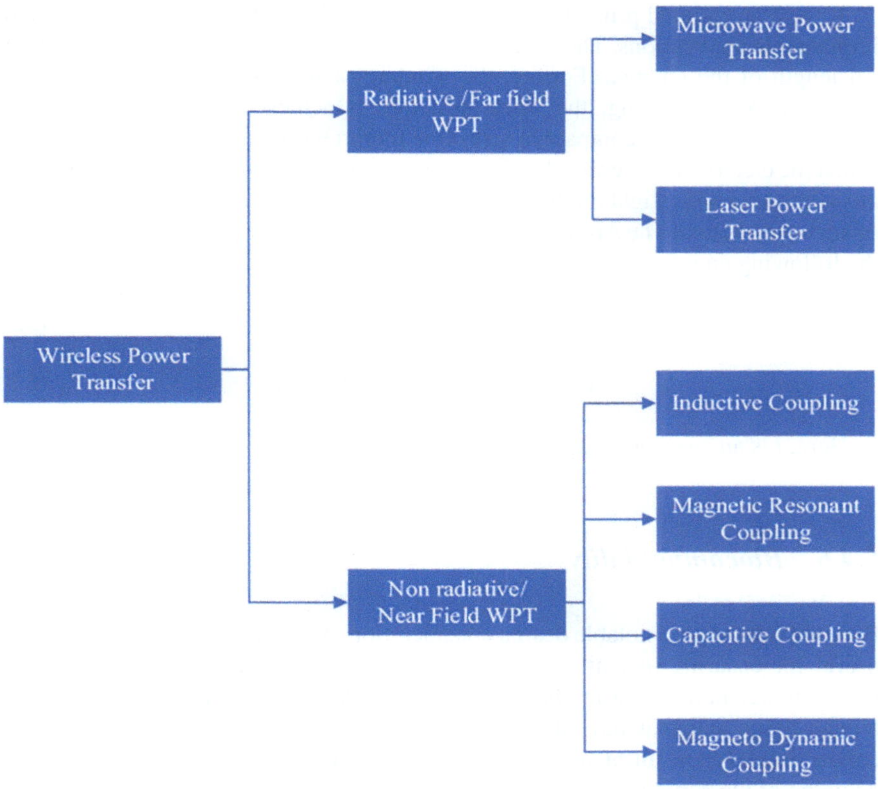

Fig. 5.1 Classification of wireless power transfer methods [12]

magnetic resonant coupling mechanisms are more popular than capacitive and magneto-dynamic coupling [13, 14]., [12].

5.4.5 Miniaturization

Another aspect that must be addressed is the design of medical devices. A trade-off has been set between the size and the performance of the device without hampering the functionality. Different miniaturization techniques are followed to reduce the size of the antenna, such as shorting pins. The reduced size of the antenna reduces the overall size of the medical devices. For biotelemetry, the size of the antenna is not a constraint. However, miniaturization is a big challenge for implantable antennas. The electric properties of the tissue are variable and depend upon the person's age, body mass index (BMI), sex, etc. An antenna with a sharp bandwidth suffers from off-tuning because of the dynamic resonance shifts and antenna performance degrades. The antenna requires sufficient power for a sustained communication link [3].

If the conventional printed antenna length is too long, then it cannot be used for implantable applications. The shorting techniques can be used to reduce the electrical length of the antenna. Further, the winding of the lengthened arm is also an approach. After winding, the spiral structure is achieved for the antenna. The antenna becomes very compact in size. One important feature of winding of length is that the electrical properties do not change significantly, and desired performances are achievable with slight modification in the antenna structure. PIFA is an example of such an antenna. The Archimedean spiral shape in the xy plane is obtained using the following equations:

$$x(t) = 0.23\cos(t)\left(20 - (t/2)\right) \tag{5.2}$$

$$y(t) = 0.23\sin(t)\left(20 - (t/2)\right) \tag{5.3}$$

Here, t is an auxiliary parameter [8].

5.4.6 Biocompatibility

Biocompatibility is a notable design consideration in implantable antennas, and it is useful for ensuring patient safety and avoiding radiation to surrounding tissues. Furthermore, human tissues have a natural tendency to conduct electricity. So, human tissue antennas have the ability to make direct contact with the antenna's conducting layer, resulting in a short circuit. This undesirable short circuit can cause detuning of the communication, or sometimes it can completely shut the communication down. Biocompatibility is addressed via two methods: the use of a biocompatible coating (encapsulation by biocompatible material) and the use of a dielectric superstrate layer [15].

The first method is used to avoid the contact of IMD with the tissue. The IMD is encapsulated with a biocompatible chemical. The biocompatible material should not only provide the coating to save the device tissue environment but also ensure that it does not affect the antenna characteristics. Teflon with a relative permittivity (ε_r) value of 2.1 is one of the preferred choices as a biocompatible material. Alumina ($\varepsilon_r \approx 9.9$) is preferred as a biocompatible material where a high ε_r is desired. A detailed list of different biocompatible materials and their common uses in implantable devices is given in Table 5.2.

If we compare the two biocompatible materials, one is Teflon, which has lower values of relative permittivity, and the other is Alumina, which has high relative permittivity. Encapsulating the antenna with Teflon and inserting it into tissue shift the operating frequency downwards compared to the antenna operating in free space and without encapsulation. With Alumina, this downward shift is more prominent. However, the axial ratio (AR) in Teflon remains in an acceptable range. In the case of Alumina, it may exceed 3 dB. The difference in AR of the abovementioned two materials is observed because of incident waves at the interface of the materials. If the relative permittivity of two materials is different, the vertical component of the transmission coefficient behaves differently from the horizontal component. The difference in the relative permittivity is evident because most of the antennas are designed with substrates such as FR4 (with relative permittivity $\varepsilon_r \approx 4.3$) and inserted in the tissues with different relative permittivity (for fat tissue, relative

Table 5.2 Biocompatible materials, their electrical properties, and applications

Material	Relative permittivity (ε_r)	Common uses in implants
Silicone rubber	2.7–3.5	Encapsulation, flexible antennas, coatings
Polydimethylsiloxane (PDMS)	2.3–2.8	Soft implant coatings, antenna substrates
Polytetrafluoroethylene (PTFE, Teflon)	2.0–2.1	Insulation, dielectric layers, bio-coatings
Polyimide (Kapton)	3.4–3.5	Flexible circuits, implant coatings, substrates
Hydroxyapatite (HA)	8–10	Bone implants, dielectric layers
Parylene-C	2.9–3.15	Biocompatible coatings, encapsulation
Zirconia (ZrO_2)	22–30	High-dielectric applications, ceramic implants
Alumina (Al_2O_3, sapphire)	9–10	Ceramic antennas, biomedical sensors
PMMA (acrylic, Plexiglass)	2.6–3.0	Optical implants, bio-encapsulation
PEEK (polyether ether ketone)	3.1–3.5	Implant structures, antenna substrates

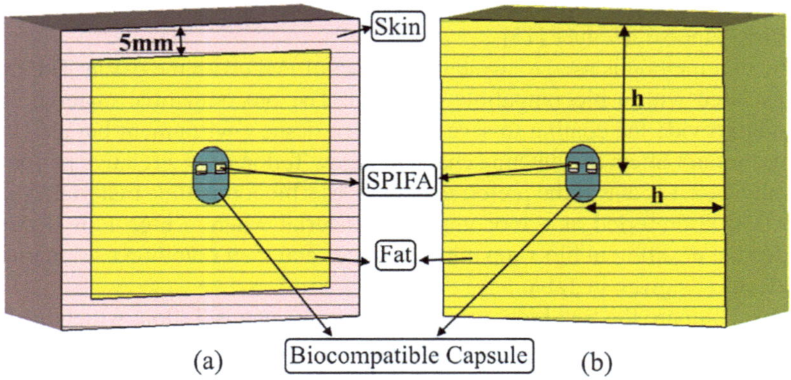

Fig. 5.2 Use of biocompatible materials in implantable antenna design on the skin [8]

Table 5.3 Commonly used superstrates in implantable devices

Commonly used superstrate materials	Electrical properties
Teflon	$\varepsilon_r = 2.1$, $\tan\delta = 0.001$
MACOR	$\varepsilon_r = 6.1$, $\tan\delta = 0.005$
Ceramic alumina	$\varepsilon_r = 9.4$, $\tan\delta = 0.006$

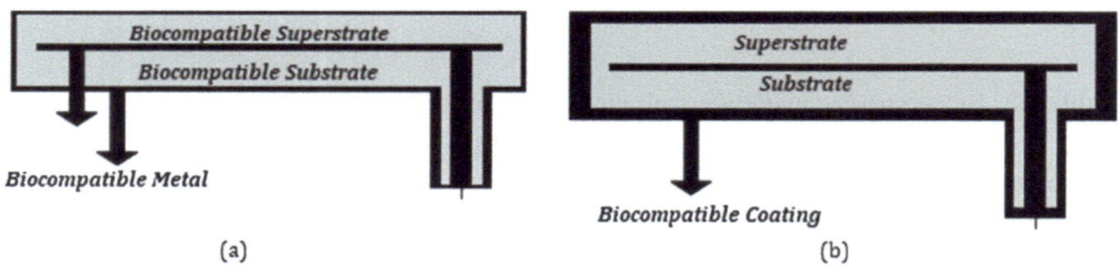

Fig. 5.3 Use of biocompatible superstrate in implantable antenna [12]

permittivity $\varepsilon_r \approx 5.5$) [8]. An example of the biocompatible capsules that are put under the skin is shown in Fig. 5.2.

The second method is the use of a dielectric superstrate layer to separate the antenna metal radiator from biological tissue and conserve the biocompatibility of the implantable antenna. The commonly used materials as a superstrate are shown in Table 5.3.

The utilization of a superstrate of a high permittivity dielectric material is another way to provide electric isolation. It also helps in achieving miniaturization as it condenses the effective wavelength. Therefore, the resonant frequency shifts downwards. A thicker superstrate increases the resonant frequency of the implant antenna. This results in larger dimensions of the IMD [12] (Fig. 5.3).

Table 5.4 Electrical properties of different tissues

Tissue type	ε_r (400 MHz)	σ (S/m, 400 MHz)
Skin	44.0	0.69
Fat	5.6	0.05
Muscle	58.5	0.80
Bone (cortical)	12.0	0.14
Bone (marrow)	6.3	0.06
Blood	68.9	1.27
Liver	56.2	0.89
Brain (gray matter)	63.4	0.98
Brain (white matter)	38.5	0.57
Lung (inflated)	22.0	0.34
Lung (deflated)	50.1	0.90
Heart	59.4	0.91

5.4.7 Proximity Effects

The communication link can be hampered by signal loss or counteract detuning if the human tissue starts interacting with the path. The electric properties of the tissues are very variable. Due to the dynamism, the antenna with low bandwidth can suffer from resonance shift, which results in unreliable radiating performance. Further, the antenna designed for one tissue environment cannot be applied to another tissue. The reason is the variation in the electric properties from tissue to tissue. The electric properties of different tissues at 400 MHz frequency are mentioned in Table 5.4. The electric properties include the conductivity and the relative permittivity. The 400 MHz frequency falls in the MICS and MedRadio bands and is, therefore, suitable for implantable devices [16, 17].

There are many methods to achieve better performance with the implant antennas. One of the coolest methods is the beam-steering method. Directional beam steering through EM switches is implemented to change the direction of radiation. Therefore, the antenna does not need to penetrate a thicker or condensed tissue, and it can switch the direction of the beam. More radiation can come out from the body, and better performance is achieved. The radiation pattern steering is achieved by using pattern reconfigurable antennas and is discussed in detail in Chap. 9.

5.4.8 Gain in the Far Field

In most implanted antenna communication systems, the off-body base station is placed in close proximity to the patient, i.e., approximately two meters from the IMD device. This off-body base station is placed to record the vital parameters and the real-time monitoring data coming from the MICS devices. This is possible with a short-range communication link. The radiation transmitted from IMD should produce sufficient gain at the receiving base station. In addition, the patient's safety is a challenge in such communication. In such a case, the effective radiated power (ERP) in the MICS band should not exceed the limit of −16 dBm (25 μW). With such low radiated power, the antenna designer has to take care of the transmission range and the SAR limits together. ERP and SAR are two constraints that limit the gain in the antenna's far field. However, for long-range telemetry communication, a higher far-field gain is desired [12].

5.4.9 Durability and Reliability

Temperature fluctuations, movement of the human body, and orientation of devices are major concerns for medical devices. The device should have an insulated coating when it is implanted inside the tissue. All these factors must be considered to ensure the antenna withstands the harsh surrounding conditions.

5.4.10 Regulatory Compliance

To ensure the safety of medical devices, some standards are set by regulatory bodies such as the IEEE, ICNIRP, FDA, and IEC (Table 5.5). The SAR is a crucial parameter for ensuring patient safety in implant devices. The SAR is defined as the rate of energy deposited by a unit mass of tissue. The maximum permissible power applied to the antenna in implant devices can control the radiation. The SAR regulations are advised by IEEE. The IEEE C 95.1_1999 standard states that the SAR averaged over 1 gram of cube-shaped tissue is less than 1.6 W/kg (SAR1g, max ≤1.6 W/kg). The International Commission on Non-Ionizing Radiation Protection (ICNIRP) has stated that SAR averaged over 10 grams of tissue should be less than 2 W/kg. If we compare both the standards, i.e., IEEE c95.1_1999 and ICNIRP for SAR, an average of over 10 grams of cube-shaped tissue is SAR10g, max ≤2 W/kg for IEEE c95.1_1999 and less than 2 W/kg for the ICNIRP [18, 19].

Table 5.5 Regulatory compliance on the SAR limits

Regulatory body	Frequency band	SAR limits	Applications
Federal Communications Commission (FCC)	MedRadio band 401–406 MHz	1.6 W/kg (averaged over 1 g of tissue)	Power limitation: ≤25 μW ERP (for implant communication devices)
ICNIRP (International & EU)	MedRadio (401–406 MHz) and ISM (2.45 GHz) bands	2.0 W/kg (averaged over 10 g of tissue)	Used by CE (Europe), WHO, and many international bodies
IEEE C95.1–2019	International standard	2 W/kg (10 g) for head and body	Applies to implantable and wearable medical devices. Defines maximum permissible exposure (MPE) limits

Further temperature rise because of prolonged use of the battery is another challenge. For example, the prolonged use of a pacemaker can raise the temperature. The continuous use of EM radiation can increase the temperature around the body. The WPT mechanisms are used for recharging the battery and can increase the temperature.

5.5 Tuning Techniques for PIFAs in Medical Devices

5.5.1 Adjustable Shorting Pins

The shorting pin connects the ground plane with the conducting strip. The position of the shorting pin is crucial to fine-tune impedance matching and achieve the desired resonant frequency for specific applications. This ensures the optimum efficiency. The shorting pin is placed between the patch and the ground plane. This results in reducing the physical dimension of the antenna and, therefore, the operating frequency. The shorting pin is a reason to increase the bandwidth in both conventional patch antennas [20].

5.5.2 Tuning of Feed Position

The feeding is used to excite the antenna for operation. The proper feed position reduces the return loss and maximizes the power transfer to the antenna. The advantage offered by position optimization in the antenna is the ability to achieve impedance matching. For impedance matching purposes, it is possible to use the loading methods where the appropriate inductance or capacitance is loaded to offset the frequency band and direction in the desired range. Further, the loading methods can work as a miniaturization technique to reduce the size of the antenna. A split-ring resonator (SRR) can also be merged to achieve impedance matching and antenna size reduction [9].

5.5.3 Slot/Notch Optimization

Tuning the antenna can be done by changing the geometry of slots/notches of the microstrip antenna. Cutting slots or notches in the conducting part of the antenna causes a change in the current orientation of the antenna. Therefore, it helps to improve radiation efficiency by enabling multiple bands for communication. The bandwidth can also be enlarged for the selected bands.

5.5.4 Material Selection

The antenna material is crucial for attaining optimum performance. The dielectric substrate with optimal permittivity is used to balance size, efficiency, and performance. For an implant antenna, the selection of antenna dielectric material is decided by the tissue under consideration. The effective dielectric and conductivity of different tissues are mentioned in the table below. The equations for the calculation of effective dielectric and conductivity are given in eqs. 5.4 and 5.5.

$$\varepsilon_{\text{eff}} = \varepsilon' - \frac{\sigma''}{\omega} \tag{5.4}$$

$$\sigma_{\text{eff}} = \sigma' + \omega\varepsilon'' \tag{5.5}$$

Here, in the equations, the ω is the $2\pi f$, and f represents the desired resonant frequency of the antenna. Further, a trade-off has to be set, as the high permittivity of the material will reduce the bandwidth and efficiency.

5.5.5 Patch Staking

In the case of a patch antenna, sometimes the multiple patches are mounted vertically. This extends the radiator's current flow path, and the overall dimensions of the antenna are decreased. A metamaterial-based superstrate is added with an antenna structure to enhance the axial ratio, bandwidth, and gain. The patch-stacking approach is not frequently used in IMD applications. As the size of the antenna becomes large, applications in MICS become limited for a large antenna [12].

5.5.6 Attaching Tuning Devices

Reconfigurable antenna design can also be implemented in medical devices. Electronic tuning devices, such as varactor diodes and PIN diodes, are attached to the conducting patch of the antenna to achieve reconfigurability. The different ON and OFF combinations of these diodes change the current circulation in the antenna. A frequency/pattern reconfigurable antenna is achieved. The details of the reconfigurable antenna are mentioned in Chap. 9.

5.5.7 Simulation Tools

The two most popular tools for electromagnetic simulation nowadays are HFSS and CST. These simulation tools have the capability to predict and optimize the antenna before prototyping.

5.6 Problem Formulation

The main objective of this chapter is to design a compact, high-efficiency Planar Inverted-F Antenna (PIFA) for reliable wireless communication in medical implantable devices. The antenna should fall in the MICS band for communication. The antenna should have applications in pacemakers, glucose monitors, and neural implants. These applications require an efficient communication system to transmit data to external devices for monitoring and diagnostics. The antenna size and biocompatibility are two significant challenges that must be addressed. Therefore, the antenna should:

- Operate within the frequency ranges suitable for **MICS** and **MedRadio** or **WMTS bands**.
- Ensure stable performance despite changes in **body tissue properties** (e.g., glucose levels in diabetic patients).
- Achieve sufficient **bandwidth**, **efficiency**, and **low Specific Absorption Rate (SAR)** for safe use in biomedical implants.

5.7 Related Work

The design of antennas for implantable medical devices has been extensively explored, focusing on achieving compactness, biocompatibility, and efficient operation across medical communication bands. Key studies include the literature survey mentioned in Table 5.6.

5.7.1 Key Insights from Literature

- Compact design and dual-band operation: Existing designs have successfully miniaturized antennas while supporting multiple frequency bands, such as MICS and WMTS.

Table 5.6 Literature review

Author and year	Key feature	Details
[21]	A dual-band PIFA is designed for wireless biotelemetry in the MICS and WMTS bands	Designed a dual-band PIFA antenna optimized for MICS and WMTS bands. Rogers RO3210 material enhances signal integrity, providing stable data transmission. This study demonstrated the potential for compact, efficient designs in biotelemetry applications
[22]	Bandwidth enhancement and size reduction techniques for implantable PIFA	Embedding hook-shaped slots in the edges of a stacked PIFA enhances bandwidth and reduces size, making it ideal for implantable biotelemetry devices. The design achieves a compact size of $8 \times 8 \times 1.9$ mm^3 and a 122 MHz −10 dB impedance bandwidth covering the MICS band. This innovation results in 144% wider bandwidth and a 36% smaller size compared to other compact antennas
[23]	PIFA is designed for in-body medical devices support	A compact, dual-frequency human tissue-implantable rectenna, combining a miniaturized PIFA and rectifier system, is designed for wireless data telemetry and power transmission in the MedRadio (402–405 MHz) and ISM (902.8–928 MHz) bands. The PIFA's size reduction and dual-band operation are achieved through slit insertion and L-shaped slot loading. The design's resonance, radiation, and SAR performance are evaluated, alongside RF-to-DC conversion efficiency, ensuring effective power transmission and safety in medical applications
[24]	Designed an implantable PIFA for general biomedical applications in the ISM band	A compact PIFA antenna operating in the 2.4 GHz ISM band is proposed for implantable medical devices, designed with copper and Rogers RO3010 materials. The antenna's small size (9.48 mm × 7.8 mm × 0.25875 mm) and safety encapsulation ensure compatibility with various implantable devices. Simulations using CST microwave studio demonstrate its effectiveness in both rectangular and cylindrical phantom models, validating its suitability for medical applications
[5]	Designed PIFA for MICS band	A compact implantable stacked PIFA antenna for biotelemetric communication in the 402–405 MHz MICS band is designed with embedded slots on each radiating patch, achieving a size reduction of 0.013 λ. the antenna, measuring $10 \times 10 \times 1.905$ mm^3, shows a − 10 dB bandwidth of 29 MHz, a radiation efficiency of 0.9%, and a far-field gain of −18.8 dB. Both in vitro and in vivo tests confirm the design's performance, with SAR calculations ensuring safety, offering a smaller and comparable alternative to existing implantable antennas for biotelemetry applications
[25]	Designed a scalp-implantable antenna for the MICS band	Proposed scalp-implantable antennas for telemetry in MICS and ISM bands. They addressed safety concerns such as SAR compliance and conducted a link budget analysis to ensure reliable communication in challenging environments

(continued)

Table 5.6 (continued)

Author and year	Key feature	Details
[26]	PIFA for brain application (MICS and WMTS bands)	This paper studies a flexible thin film for wireless brain interface applications for brain–machine interface microsystems, including brain–computer interfaces (BCIs). PIFA operates at two frequency bands, the MICS and WMTS bands. The antenna structure is designed to achieve power efficiency without violating SAR limits. Flexible material DuPont Kapton with a thickness of 0.25 mm is used, and the silicon is used as an adhesive superstrate. Further, a prototype is built and tested in a brain-mimicking solution. The results demonstrate that the prototype successfully provides an acceptable result between simulation and measurement
[27]	Tri-band implantable antennas	Proposed a tri-band implantable antenna for wireless telemetry. Their design optimized size and operational efficiency across multiple bands, addressing the need for versatility in implantable medical devices
[28]	A quad-band antenna is designed for WMTS and ISM bands	Designed an S-shaped quad-band antenna for MedRadio, WMTS, and ISM applications. Their work demonstrated the feasibility of supporting multiple bands in a compact form factor, expanding the range of medical telemetry possibilities
[29]	Compact dual-band biocompatible antennas	Introduced a small, dual-band implantable antenna for medical telemetry applications. This design incorporated biocompatible materials and achieved efficient operation in both the MICS and WMTS bands
[30]	Implantable microstrip antennas	Developed microstrip antennas tailored for communication with medical implants. Their work emphasized achieving effective impedance matching to reduce signal losses, a critical factor in implantable devices

- Biocompatibility: Materials like Rogers RO3010 and copper have proven effective in ensuring compatibility and safety in biomedical environments.
- Signal integrity and SAR compliance: Several studies emphasize achieving low SAR values and maintaining strong signal integrity across diverse tissue environments.
- Challenges in bandwidth and efficiency: Many designs focus on enhancing bandwidth or efficiency independently, highlighting a need for balanced optimization.

5.8 Simulation Results and Discussion

The PIFA antenna is designed by a complementary meandering line in the radiating patch with a shorting ground plane. The proposed antenna is shown in Fig. 5.4. The proposed PIFA antenna design process is given in Fig. 5.5. The final antenna

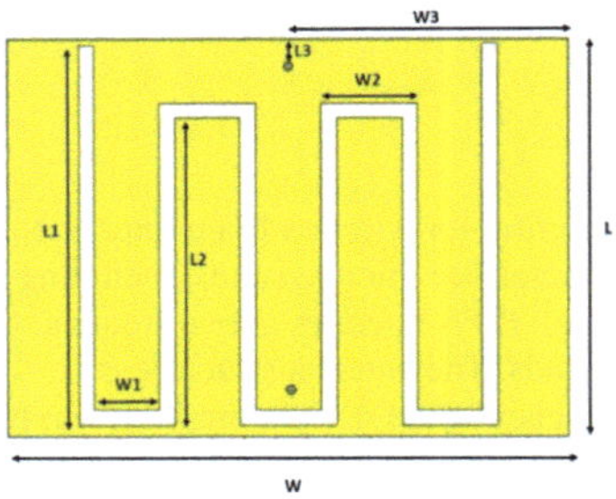

(a)

PARAMETER	SIZE (mm)
L	26
L1	24.81
L2	20
L3	1.5
W	35.61
W1	4.13
W2	6.13
W3	17.8

(b)

Fig. 5.4 (**a**) Proposed PIFA antenna and (**b**) antenna dimensions

(a) (b) (c)

Fig. 5.5 Propose antenna design process

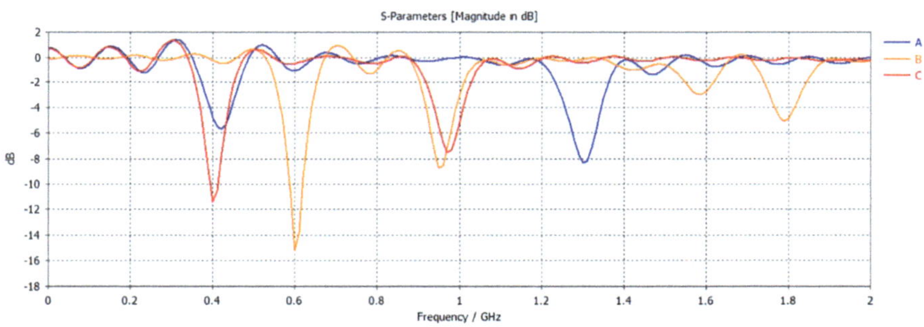

Fig. 5.6 Return loss of the proposed antenna design process

resonates at 400 MHz and is suitable for medical applications. The resonant behaviors of the antenna are justified by the current distribution diagram. The current distribution and proposed antenna return loss are shown in Figs. 5.6, 5.7, and 5.8. The antenna provides 0.00230286 W/Kg SAR at 400 MHz resonance frequency. The SAR is measured at a distance of 2.89 mm from the body. This SAR antenna is suitable for medical applications. The 2D radiation pattern (E plane and H plane) is given in Figs. 5.9 and 5.10.

Fig. 5.7 Current distribution of proposed antenna at 400 MHz

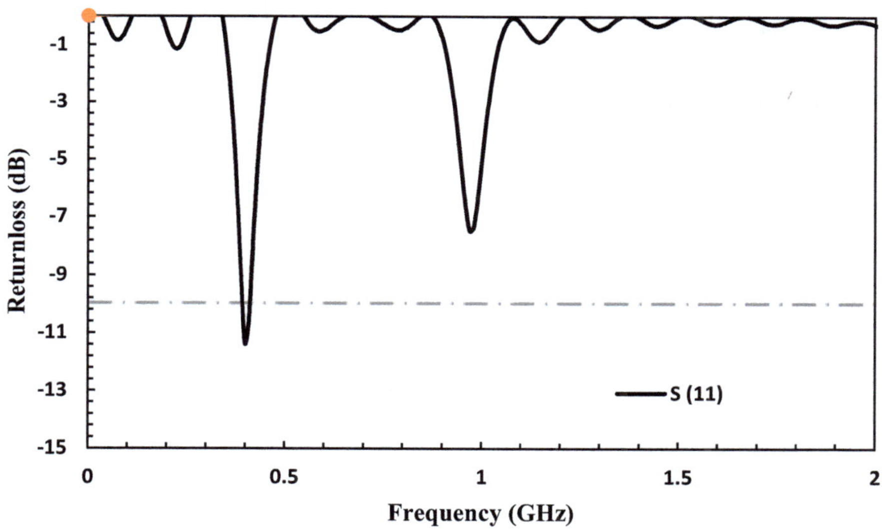

Fig. 5.8 Return loss of proposed antenna at 400 MHz

5.9 Open Research Challenges

Though the PIFA is a miniaturized antenna, there is a need for further miniaturization without compromising performance. The radiation performance should remain intact, though a reduction in antenna size is desired. The power consumption of the implant device is also crucial. Since these devices have to interact with human tissues, biocompatibility of the implant devices is another major challenge. The

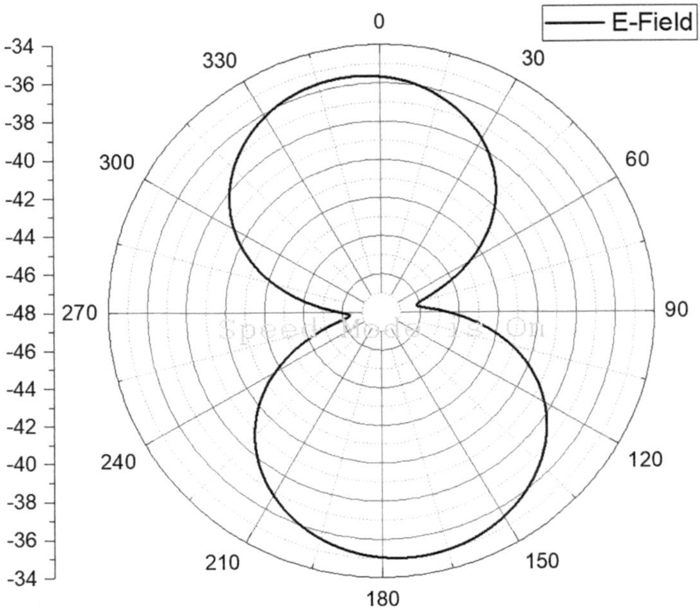

Fig. 5.9 2D E-field radiation pattern at 400 MHz

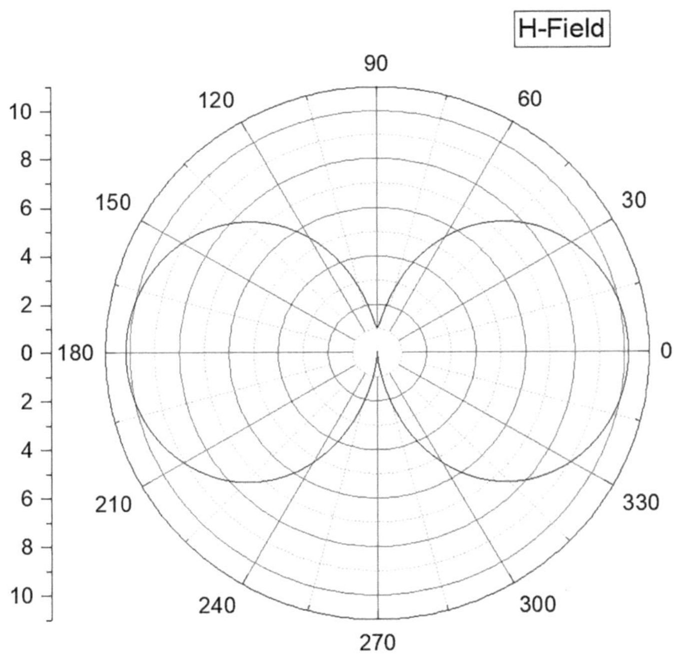

Fig. 5.10 2D H-field radiation pattern at 400 MHz

challenge is posed because the electromagnetic field penetrates through the tissue. Further research is required to minimize the SAR of the implant antennas. The challenge becomes vital if a high data rate transmission is required from the implant device. Therefore, an optimized antenna design is desired to minimize the adverse effect on human tissue. The implant devices desire communication band-specific transmission. The implanted device should radiate specifically in the MICS (401–406 MHz) band, while the WMTS band is desired for medical telemetry. Achieving such precise frequency bands while keeping antenna size small is a significant challenge and requires antenna design optimizations. The last but very significant challenge is power management and energy harvesting. Since the device has to be inside the body, its power consumption should be minimal. Much research is going on in the field of low-power operations of antennas. Further energy harvesting methods for PIFA antennas are essential to extend the lifespan of medical implants and reduce the need for frequent battery replacements.

Further, some research challenges are available for all the tiny implant antennas. The following are the essential criteria for unending research on implantable antennas:

- To uncover new materials for encapsulation and coatings that can enhance biocompatibility. The material should also minimize the chances of rejection or infection.
- To improve signal strength so that after penetrating the human tissue, the signal should reach a longer distance.
- Energy harvesting and wireless power transfer methods should be improved so that the longevity of implantable devices without the need for frequent battery replacements is ensured.
- Security and privacy of monitored data are one of the recent challenges that have to be taken care of by the researchers and the governing bodies. With the advancements in technology, the safeguarding of the patient's critical health information has opened new areas of research.
- Integration of these IMDs with emerging technologies is another challenge and needs to be taken care of. Bioelectronics and nanotechnology can lead to the development of more sophisticated and versatile medical devices.
- Customizable antennas tailored to individual patient needs and specific medical conditions must be developed to increase the effectiveness of implantable medical devices [12].

5.10 Summary

The paper explores the design and application of Planar Inverted-F Antennas (PIFA) in wireless medical devices, emphasizing their compact size, low Specific Absorption Rate (SAR), and omnidirectional radiation pattern. PIFAs are highlighted for their integration in implantable and wearable devices, ensuring reliable communication and minimal energy consumption. Challenges such as limited bandwidth, detuning in proximity to human tissue, and the need for precise optimization in medical frequency bands are addressed. The study reviews advancements in dual-band and miniaturized PIFA designs, emphasizing biocompatibility and efficient operation. Future research focuses on further miniaturization, power management, and SAR reduction to enhance their suitability for biomedical applications.

References

1. J. Shang, Y. Yu, L.-J. Xu, Compact dual-band implantable planar inverted-F antenna with bandwidth enhancement. Microw Opt Technol Lett. **62**, 322–328 (2020). https://doi.org/10.1002/mop.32009
2. M. Jasim et al., An extensive review on implantable antennas for biomedical applications: Health considerations, geometries, fabrication techniques, and challenges. Alex Eng J **112**, 110–139 (2025)
3. M.S. Zulkefli et al., Experimental wireless link and SAR assessments of an implantable PIFA for biotelemetry in the 2.45 GHz band. IEEE J Electromagn RF Microw Med Biol **7**(3), 281–289 (2023). https://doi.org/10.1109/JERM.2023.3294707
4. J. Kim, Y. Rahmat-Samii, Planar inverted-F antennas on implantable medical devices: Meandered type versus spiral type. Microw. Opt. Technol. Lett. **48**, 567–572 (2006). https://doi.org/10.1002/mop.21409
5. Z. Huang, H. Wu, S.S. Mahmoud, Q. Fang, Design of a novel compact MICS band PIFA antenna for implantable biotelemetry applications. Sensors **22**(21), 8182 (2022). https://doi.org/10.3390/s22218182
6. F. Merli, L. Bolomey, J.-F. Zurcher, G. Corradini, E. Meurville, A.K. Skrivervik, Design, realization and measurements of a miniature antenna for implantable wireless communication systems. IEEE Trans Antennas Propag **59**(10), 3544–3555 (2011). https://doi.org/10.1109/TAP.2011.2163763
7. J. Kim, Y. Rahmat-Samii, Planar inverted-F antennas on implantable medical devices: Measured type versus spiral type. Microwave Opt Technol Lett **48**, 563–572 (2006)
8. A. Abdi, F. Ghorbani, H. Aliakbarian, T.K. Geok, S.K.A. Rahim, P.J. Soh, Electrically small spiral PIFA for deep implantable devices. IEEE Access **8**, 158459–158474 (2020). https://doi.org/10.1109/ACCESS.2020.3018882
9. O.A. Iupikov et al., A dual-fed PIFA antenna element with nonsymmetric impedance matrix for high-efficiency Doherty transmitters: Integrated design and OTA-characterization. IEEE Trans Antennas Propag **68**(1), 21–32 (2020). https://doi.org/10.1109/TAP.2019.2938738
10. M. Huang et al., Highly integrated PA-PIFA with a wide frequency tuning range. IEEE Antennas Wirel Propag Lett **20**(8), 1433–1437 (2021). https://doi.org/10.1109/LAWP.2021.3085937
11. S. Rao et al., Miniature implantable and wearable on-body antennas: Towards the new era of wireless body-centric systems [antenna applications corner]. IEEE Antennas Propag. Mag. **56**(1), 271–291 (2014)
12. A. Mohan, N. Kumar, Implantable antennas for biomedical applications: A systematic review. BioMed Eng OnLine **23**, 87 (2024). https://doi.org/10.1186/s12938-024-01277-1
13. M. Ghovanloo, K. Najafi, A compact large-range wireless powering system for biomedical implants. IEEE Trans Biomed Eng **52**(10), 1635–1643 (2005)
14. J.S. Ho, A.J. Yeh, M.M. Maharbiz, Wireless power transfer to deep-tissue microimplants. Proc Natl Acad Sci U S A. **111**(22), 7974–7979 (2014)
15. R. Kumar, S. Singh, CPW-fed conformal PIFA design for implantable IoMT devices with wideband performance. IEEE Sens J **24**(1), 231–237 (2024). https://doi.org/10.1109/JSEN.2023.3329416

16. C. Gabriel, Compilation of the dielectric properties of body tissues at RF and microwave frequencies. Brooks Air Force Technical Report AL/OE-TR-1996-0037, San Antonio, TX, USA, 1996.
17. M. Maleki, M. Sarrafzadeh, and W. J. Kaiser, Design considerations for implantable PIFA antennas in biomedical telemetry applications. Proceedings of the 2019 IEEE Antennas and Propagation Symposium (APS), Atlanta, GA, USA, 2019, pp. 3342–3345.
18. S.I. Kwak, D.-U. Sim, J.H. Kwon, Y.J. Yoon, Design of PIFA with metamaterials for body-SAR reduction in wearable applications. IEEE Trans Electromagn Compat **59**(1), 297–300 (2017). https://doi.org/10.1109/TEMC.2016.2593493
19. K. Guido, A. Kiourti, Wireless wearables and implants: a dosimetry review. Bioelectromagnetics **41**(1), 3–20 (2020). https://doi.org/10.1002/bem.22240. Epub 2019 Dec 18. PMID: 31854006; PMCID: PMC9291027
20. N.C. Karmakar, Shorting strap tunable stacked patch PIFA. IEEE Trans Antennas Propag **52**(11), 2877–2884 (2004). https://doi.org/10.1109/TAP.2004.835124
21. P. Soontornpipit and P. Satitvipawee, Design and development of a dual-band PIFA antenna for wireless biotelemetry applications. 2018 International Electrical Engineering Congress (iEECON), Krabi, Thailand, 2018, pp. 1–4, doi: https://doi.org/10.1109/IEECON.2018.8712165.
22. W.-C. Liu, S.-H. Chen, C.-M. Wu, Bandwidth enhancement and size reduction of an implantable PIFA antenna for biotelemetry devices. Microw Opt Technol Lett (2009). https://doi.org/10.1002/mop.24142
23. S. Bakogianni and S. Koulouridis, Design of a novel miniature implantable rectenna for in-body medical devices power support. 2016 10th European Conference on Antennas and Propagation (EuCAP), Davos, Switzerland, 2016, pp. 1–5, doi: https://doi.org/10.1109/EuCAP.2016.7481970.
24. M. S. Ahmad, M. M. Rahman and K. M. Morshed, Design and performance analysis of a compact implantable PIFA antenna for biomedical applications. 2019 5th International Conference on Advances in Electrical Engineering (ICAEE), Dhaka, Bangladesh, 2019, pp. 617–622, doi: https://doi.org/10.1109/ICAEE48663.2019.8975528.
25. A. Kiourti, K.S. Nikita, Miniature scalp-implantable antennas for telemetry in the MICS and ISM bands: Design, safety considerations and link budget analysis. IEEE Trans Antennas Propag **60**(8), 3568–3575 (2012). https://doi.org/10.1109/TAP.2012.2201078
26. P. Soontornpipit, Design and development of a dual-band PIFA antenna for brain interface applications. 2019 7th International Electrical Engineering Congress (iEECON), Hua Hin, Thailand, 2019, pp. 1–4, doi: https://doi.org/10.1109/iEECON45304.2019.8939014.
27. M. Shakib, M. Moghavvemi, W. Mahadi and M. Ahmed, Design of a tri-band implantable antenna for wireless telemetry applications. 2014 IEEE MTT-S International Microwave Workshop Series on RF and Wireless Technologies for Biomedical and Healthcare Applications (IMWS-Bio2014), London, UK, 2014, pp. 1–3, doi: https://doi.org/10.1109/IMWS-BIO.2014.7032451.
28. C.-K. Wu, T.-F. Chien, C.-L. Yang, C.-H. Luo, Design of novel S-shaped quad-band antenna for MedRadio/WMTS/ISM implantable biotelemetry applications. Int J Antennas Propag (2012)
29. S. Seran, T. Karacolak and J. P. Donohoe, A small implantable dual band biocompatible antenna for medical wireless telemetry applications. 2013 USNC-URSI Radio Science Meeting (Joint with AP-S Symposium), Lake Buena Vista, FL, USA, 2013, pp. 212, doi: https://doi.org/10.1109/USNC-URSI.2013.6715518.
30. P. Soontornpipit, C.M. Furse, Y.C. Chung, Design of implantable microstrip antenna for communication with medical implants. IEEE Microw Theory Tech **52**, 1944–1951 (2004)

Chapter 6
Implantable Antennas for Medical Devices

Contents

6.1 Introduction

Real-time data transmission, therapeutic treatments, and ongoing patient monitoring are made possible by implantable medical devices (IMDs), which are the result of the development of wireless healthcare technology. Here, devices such as pacemakers, neural stimulators, glucose monitors, and capsule endoscopes require reliable and efficient wireless communication to transmit physiological data to external monitoring systems [1–3]. Implantable antennas play a critical role in ensuring seamless connectivity while operating within the human body, where factors such as tissue absorption, miniaturization constraints, and biocompatibility present

© The Author(s), under exclusive license to Springer Nature Switzerland AG 2026
A. Segun Ayokunle et al., *Biomedical Antenna Technology*,
https://doi.org/10.1007/978-3-031-95268-5_6

significant challenges [4–6]. The design of these antennas must address issues like signal attenuation in biological tissues, low-power consumption, and stable performance in dynamic body environments [7–10].

Implantable antennas are typically designed to operate in medical frequency bands such as the Medical Implant Communication Service (MICS, 402–405 MHz), the Industrial, Scientific, and Medical (ISM) band (2.4 GHz, 5 GHz), and Ultra-Wideband (UWB, 3.1–10.6 GHz), depending on the application [11–13]. The choice of biocompatible materials (e.g., Parylene, PDMS, or Rogers substrates) and miniaturization techniques (e.g., meandered designs, fractal geometries, or metamaterials) is crucial for ensuring both safety and performance [14–16]. As the demand for minimally invasive, long-lasting, and highly efficient IMDs grows, the development of innovative implantable antenna technologies is becoming a key focus in biomedical engineering, enabling enhanced telemedicine, personalized healthcare, and remote patient management [17–20].

Monitoring in real time: continuous monitoring of physiological data and vital signs is made feasible by implantable antennas, which may help identify medical issues early and take appropriate action. Better mobility: with implantable antennas enabling wireless connection, patients can move freely without being attached to monitoring devices, increasing their quality of life. Care is only minimally intrusive. These antennas eliminate the need for regular doctor visits and intrusive tests, making operations less invasive and more comfortable for patients [21, 22]. In addition, they reduce the possibility of infection. Remote healthcare and telemedicine: implantable antennas enable telemedicine and remote patient monitoring, making healthcare more accessible to people living in remote or underdeveloped locations. Implantable smart devices: implantable antennas are required for smart implants such as pacemakers, insulin pumps, and neurological stimulators in order to provide accurate control and fine-tuning of the devices [23–27]. Figure 6.1 shows the

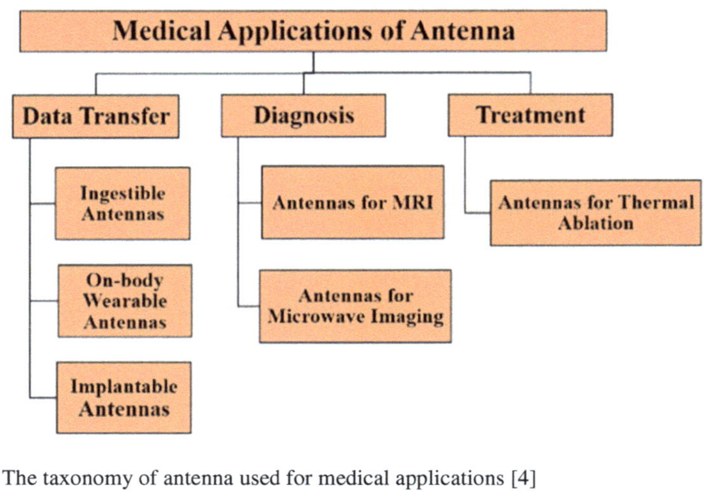

Fig. 6.1 The taxonomy of antenna used for medical applications [4]

medical application of antenna for different purposes. The antenna can be used for data transfer, diagnosis, and treatment purposes.

6.2 Challenges and Considerations for Implantable Antennas

Implantable antennas, which are fixed to various body parts, must be aesthetically pleasing in a range of settings, unlike old-fashioned antennas that function in free space. The implantable antenna should meet the following specifications: low-power consumption, far-field gain, patient safety, miniaturization, and biocompatibility. Biocompatibility is the primary design factor for the implantable antenna in order to protect the patient and prevent radiation exposure to nearby tissues. Additionally, human tissue antennas can come into direct touch with the antenna metallization layer, which could result in a short circuit, because human tissues are electrically conductive by nature. For the long-term implantation of the implantable antenna, biocompatibility and the prediction of unwanted short circuits are particularly important. A dielectric superstrate layer and a biocompatible coating are the two approaches used to address biocompatibility. The most popular method for keeping the implantable antenna biocompatible is used to utilize a dielectric superstrate layer to keep the metal radiator of the antenna away from living tissue.

One of the most crucial design factors for implanted antennas is miniaturization. Implantable antennas that are incredibly tiny are the result of recent advancements in antenna fabrication technology. The implanted devices, usually in the MICS or MedRadio bands, are assigned to the geometries of standard half-wavelength or quarter-wavelength antennas at the wave bands. Miniaturization is difficult in this state. The following is a list of some miniaturization techniques. Using a substrate/superstrate composed of a dielectric material with a high permittivity is one of the coolest methods for downsizing since it shifts high resonant frequencies to lower resonant frequencies and condenses the effective wavelength. To improve the resonance, bigger physical dimensions are needed, and a thicker superstrate raises the implant's working frequency. Radiating patches placed vertically reduce the antenna's size by lengthening the radiator's current flow channel. Antenna construction is combined with a metamaterial-based superstrate to improve the axial ratio, bandwidth, and gain.

The proper capacitance or inductance is utilized to offset the frequency band and direction to accomplish impedance matching using loading techniques. The antenna size is reduced by using both capacitive and inductive loadings. Moreover, impedance matching and size reduction can be accomplished by merging the split-ring resonator (SRR). Radiation efficiency, which gauges how well an antenna transforms input power into radiated electromagnetic energy, is an important consideration for implanted antennas. It is challenging to achieve high radiation efficiency with implantable antennas, nevertheless, because biological tissues have a tendency to absorb and attenuate electromagnetic signals. More than one frequency-dependent mechanism affects the effectiveness of EM energy transfer from dispersive tissues to free space. Lossy media include physical limitations on the radiation efficiency of electrically tiny sources, reflection (mismatch) losses due to impedance contrast, and attenuation from dielectric and conductive losses. Electrically tiny antennas can be made more radiation efficient while still being compact by reducing bandwidth and applying dielectric loading to the antenna construction, utilizing a superstrate and a high-permittivity substrate.

According to IEEE rules and recommendations used to prevent patient health, the specific absorption rate (SAR) parameter—the rate of energy deposited per unit mass of tissue—is used to measure patient safety. The maximum amount of power that can be applied to the implanted antenna is connected to patient safety. An electromagnetic field raises the temperature of bodily tissue. The implantable antenna's surrounding tissue temperature is not. The temperature rose in certain pacemaker and rectenna applications due to extended battery charging and rectifier (wireless power transfer) circuits. To establish a reliable biotelemetry link, SAR limits the receiver antenna's ability to detect the far-field gain in addition to effective radiated power. However, in order to extend the biotelemetry communication link distance, an implantable antenna with a higher gain is required. Low-power consumption limits the functionality of implanted medical devices, and there are a number of methods for charging batteries, including the inductive-loop technique. Biotelemetry links, however, are only utilized when the required data must be sent.

Due to battery lifetime and capacity limits, wireless power transfer (WPT) techniques are used in biomedical sensors and medical implants. They are also used in various biomedical applications, including retinal prostheses, pacemakers, brain stimulators, and capsule endoscopes. The difficulties associated with replacing surgical batteries are reduced via wireless power transmission, which provides a secure and easy way to overcome power capacity restrictions. Many biomedical implants have been employed for WPT recently, such as heart rate monitors, brain simulators, left ventricular assist devices, and capsule endoscopes [23–27].

6.3 Biocompatible Materials for Implantable Antennas

Biocompatibility is the primary design factor for the implantable antenna to protect the patient and prevent radiation exposure to nearby tissues. Additionally, human tissue antennas can come into direct contact with the antenna metallization layer, which could result in a short circuit, because human tissues are electrically conductive by nature. For the long-term implantation of the implantable antenna, biocompatibility and the prediction of unwanted short circuits are particularly important. A dielectric superstrate layer and a biocompatible coating are the two approaches used to address biocompatibility. The most popular method for keeping the implantable antenna biocompatible is to utilize a dielectric superstrate layer to keep the metal radiator of the antenna away from living tissue. Commonly used superstrate materials include Teflon, MACOR, and ceramic alumina. Another approach is to shield the implantable material with a thin layer of biocompatible materials such as zirconia, PEEK, and silastic MDX 4210 [23–25].

6.3.1 *Frequency Range: MBAN (Medical Body Area Network) Bands*

Radio frequency allocation for medical implants varies by region and regulatory body. The Federal Communication Commission (FCC) divides frequency band distribution in the United States into short- and long-range bands. In Europe, spectrum allocation is divided between active medical implants and associated peripherals, and the European Communication Committee (ECC) governs the medical data acquisition area. Building an antenna for use in the body is a more complex undertaking than building an antenna for open space. The frequency bands listed below employed in biotelemetry applications are:

(1) Medical implant communication services (MICSs): 402–405 MHz; (2) Medical device radio communications service (MedRadio): 401–406, 413–419, 426–432, 438–444, and 451–457 MHz; (3)Industrial scientific and medical (ISM) bands: 433.05–434.79, 902–928, 2400–2483.5, 5725–5850 MHz; (4)WMTS bands (608–614 MHz, 1395–1400 MHz, 1427–1432 MHz; (5) ultra-wideband (UWB): 3100–10,600 MHz, and medical body area network (MBAN): 2360–2400 MHz. Higher working frequencies keep narrow wavelengths, reducing antenna size. For improved data connection, antennas operating at higher frequency range with broad bandwidths are required, but they are influenced by a bigger drop in the amount of tissue compared to the tiny frequency ranges. Figure 6.2 shows an implantable antenna for injectable and ingestible purposes. It can be used for off-body, on-body, and in-body applications.

Every 2 cm of physiological material resulted in a 20–30 dB attenuation in the frequency range of 3–5 GHz. The lower operating frequencies limit communication speed, as does the usage of large dimension antennas and heavy circuit elements, which increase the size of the implantable medical device. In this image, some

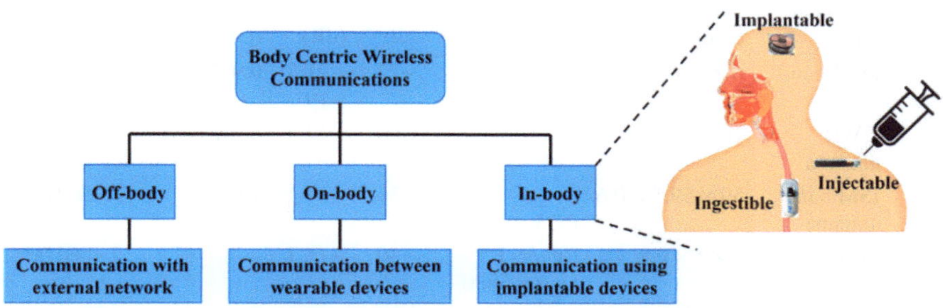

Fig. 6.2 Taxonomy of body-centric wireless communication [9]

implantable medical devices are positioned correctly on various bodily sections. Data is collected from the IMD, accessible via a cloud database, and transmitted to the appropriate individuals via wireless media [26, 27].

6.4 Problem Formulation

To identify innovative materials and coatings that can improve biocompatibility while reducing rejection or infection, to enhance signal strength and dependability in the complex and varied human body, and to assure the lifespan of implantable devices that do not require regular battery replacements, effective ways for gathering energy and wirelessly delivering it must be devised. As implantable devices grow more networked, ensuring the security and privacy of transmitted data becomes increasingly important. The combination of implantable antennas and other developing technologies, such as bioelectronics and nanotechnology, has the potential to result in more sophisticated and adaptable medical devices. To improve the efficacy of implantable medical devices, customizable antennas tailored to individual patient demands and specific medical situations must be created.

6.5 Related Work

Researchers have focused on designing compact antennas to fit within small medical implants. Techniques like meandered structures, spiral designs, and high dielectric substrates are widely used. Some previous work on implantable antennas and their performance are tabulated in Table 6.1.

Table 6.1 Previous relevant implantable antenna and their performance

Ref	Antenna type	Size	Substrate	S11 (dB)	Gain (dB)	Bandwidth (GHz)	Resonance frequency
1	Dual-band antenna	6 × 6 × 0.25	Rogers RO3010	−22	−21.8 −19.2	0.9–2.9	915 MHz 2.4GHz
2	UWB antenna	10 × 9 × 0.7	Taconic TRF 43	−40	−19	3–14	–
3	ISM band antenna	50 × 59.5 × 0.6	Rogers RO3010	−36	−14	1.9–2.6	2.45GHz
4	Wideband antenna	8 × 8 × 0.8	Rogers RT Duroid5880	−40	−16.7	–	2.45GHz
**	Proposed antenna	10 × 10 × 0.8	Rogers RT Duroid 5880	−37	<−0.1	5–7.5	6GHz

6.6 Simulation Results and Discussion

The proposed design of the antenna uses CST Studio. Figure 6.3 shows the geometric configuration of the antenna. Figures 6.3a–c show the top view, bottom view, and perspective view of the proposed single element antenna, respectively. The element with microstrip line on RT Duroid 5880 substrate has a dielectric constant of 2.2, a tan value of 0.0009, a thickness of 0.8 mm, and partial ground. The overall dimension of the single element is 10 mm × 10 mm × 0.8 mm.

Figure 6.4 shows its reflection coefficient (S11) having impedance bandwidth ranges from 5 GHz to 7.5 GHz. The upper surface of the antenna forms a shape radiator excited via a 50-ohm microstrip feed line and partial ground plane acting as defective ground surface (DGS). DGS acts as a reflector to disturb the flow of surface current distribution on the ground plane with increased isolation and shift to impedance bandwidth range within its resonant band expands. DGS has a high current density that is integrated into the radiator to improve impedance matching and bandwidth area. Impedance matching and band resonance behavior are affected by variations in the dimensions of DGS.

6.6.1 Parametric Analysis: Variation of the Width of Ground and Width of Microstrip Feed Line

The width of the ground is noted as "W1." Varying the ground width improves the impedance matching has been shown in Fig. 6.5. The length varies from 2 mm to 0.5 mm. In response, (S11) also varies from −5 dB to −37 dB to get a better result at 0.5 mm width. Similarly, the width of the microstrip feed line is indicated as "L2." The variation of "W3" from 0.4 mm to 0.6 mm with respect to varies the

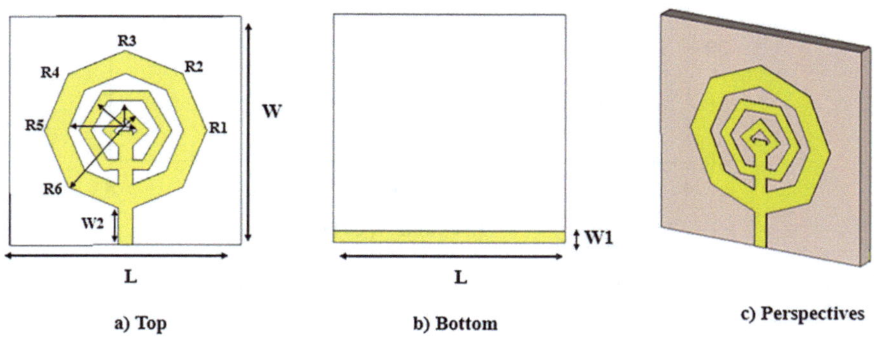

Fig. 6.3 Structural layout of antenna for design: (**a**) top view, (**b**) bottom view, and (**c**) perspective view. Dimensions: L = W = 10 mm, W1 = 0.5 mm, L2 = 0.6 mm, W2 = 5 mm, R1 = 0.5 mm, R2 = 1 mm, R3 = 1.5 mm, R4 = 2 mm, R5 = 2.5 mm, R6 = 3.5 mm

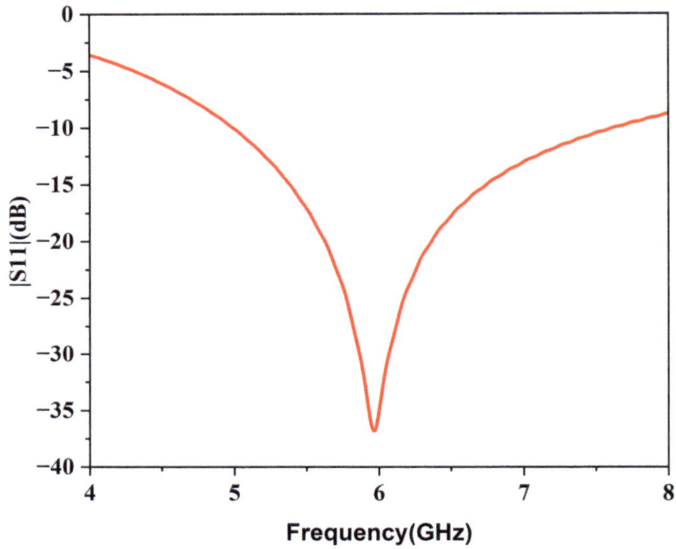

Fig. 6.4 S-parameter performance for proposed antenna design

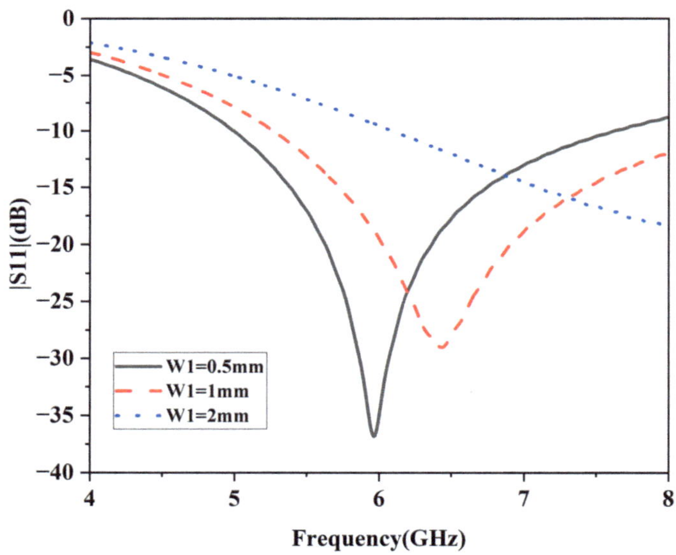

Fig. 6.5 S-parameter performance by changing W1 for proposed antenna design

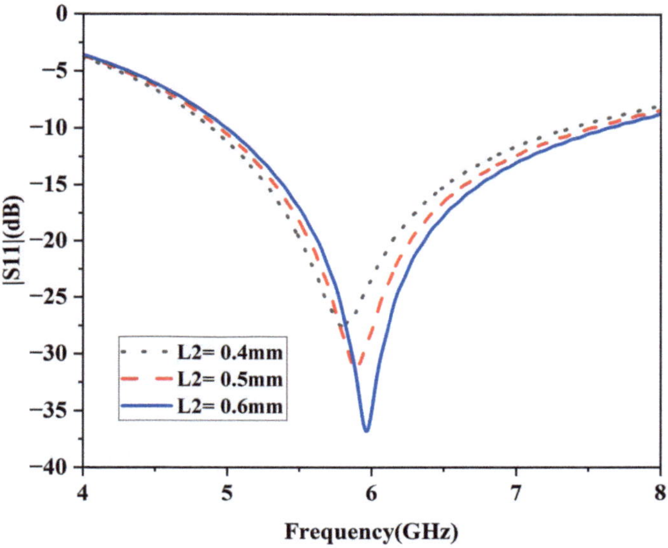

Fig. 6.6 S-parameter performance by changing L2 for proposed antenna design

bandwidth range and resonance frequency as seen in Fig. 6.5, finally fix of L2 (0.6 mm) has an impact on the antenna bandwidth to generate proper band width range (5–7.5 GHz) and good resonance frequency range (−37 dB) (Fig. 6.6).

6.6.2 Radiation Pattern

Radiation patterns are the polarization and magnitude of the EM field radiated by a transmitting antenna as a function of space coordinates. The far-field radiation pattern of the proposed antenna was simulated at frequencies (5.5 GHz, 5.87 GHz, 6 GHz) throughout whole operational bands. Figure 6.7 shows far field and 3D radiation pattern at frequency 5.5 GHz. It has directional radiation patterns that are most stable in this band, depending on the position of the antenna. From all directions, the antenna accepts signals and radiates omnidirectional. Figures 6.8 and 6.9 show far-field and 3D radiation pattern at frequency 5.87 GHz and 6 GHz, respectively.

6.6.3 Surface Current Distribution of Antenna

Figure 6.10 shows the surface current distribution of the proposed antenna at a resonant frequency of 5.87 GHz. It is observed from the figure, the surface current concentration is higher at the shape radiator, which shows it is the main contributor

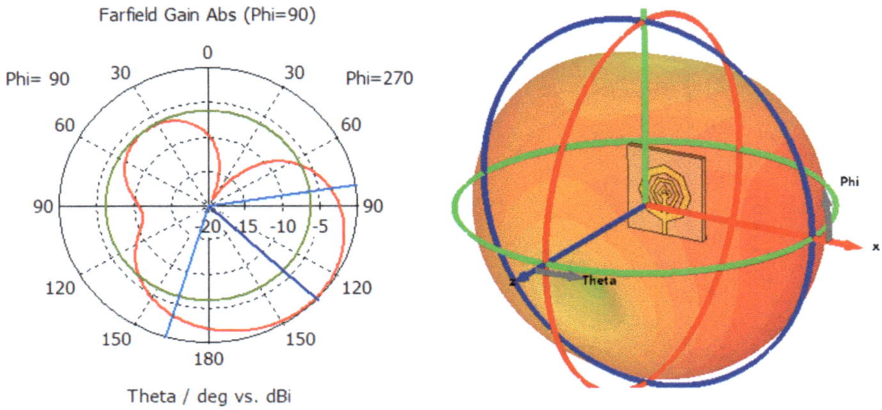

Fig. 6.7 Far-field and 3D radiation pattern at frequency 5.5 GHz for proposed antenna design

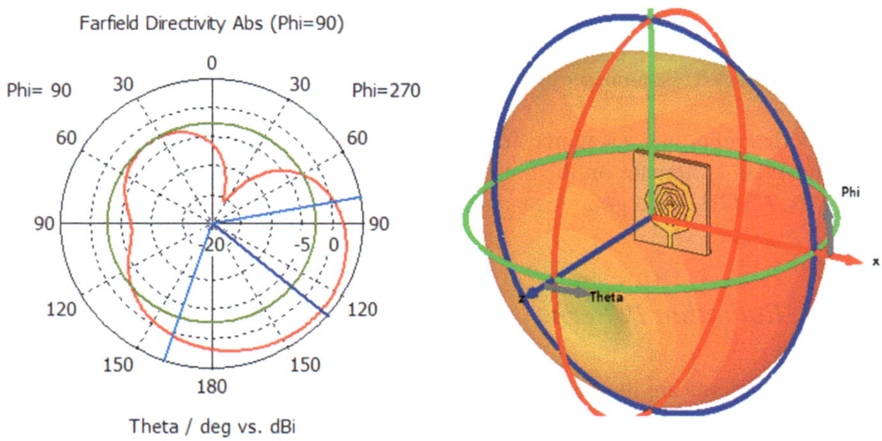

Fig. 6.8 Far-field and 3D radiation pattern at frequency 5.87 GHz for proposed antenna design

to the resonance. In order to investigate the EM interaction between the elements, the surface current distribution is recorded for various phases of the input excitation signal such as 0°, 90°, 180°, and 270°.

6.6.4 Gain

An antenna transmits or receives EM waves in a given direction. The EM waves radiated or received by an isotropic antenna are called gain. Figure 6.11a shows the simulated value of gain at the resonance frequencies 6 GHz. Figure 6.11b depicts the VSWR value against frequency.

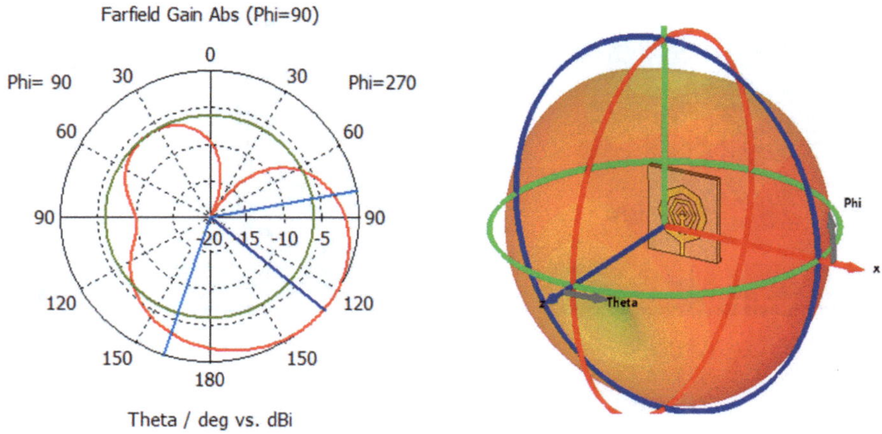

Fig. 6.9 Far-field and 3D radiation pattern at frequency 6 GHz for proposed antenna design

6.6.5 VSWR

The ratio of the maximum voltage to the minimum voltage in a standing wave is known as the "Voltage Standing Wave Ratio." The term, which indicates the impedance mismatch, is VSWR. The higher the impedance mismatch, the higher the value of VSWR. The ideal value of VSWR should be 1:1 for effective radiation, as shown in Fig. 6.11b.

6.6.6 SAR in Phantom

In biomedical applications, an implantable antenna collects and transmits images and videos of internal organs in the GI tract to an external receiver. This data is sent via radiofrequency impulses that flow through the human body. When human tissue is subjected to information-carrying signals, it may heat up. If the heat produced by these signals surpasses certain thresholds, it becomes unsuitable for human health. Limits are usually described in terms of specific absorption rate (SAR), which assesses the effect of electromagnetic radiation on tissues. Low SAR values are critical to preventing tissue heating and consequent injury. Maintaining safe levels of SAR during prolonged exposure within the human body is crucial. According to IEEE safety requirements, SAR should not exceed 1.6 W/kg over 1 g of tissue or 2 W/kg over 10 g of tissue with a 1 W input power. Following these rules decreases the likelihood of damaging heat impacts on tissues while also ensuring that the devices perform safely inside the human body. Figure 6.12 shows the SAR (1 g) distribution of the proposed antenna at 6 GHz.

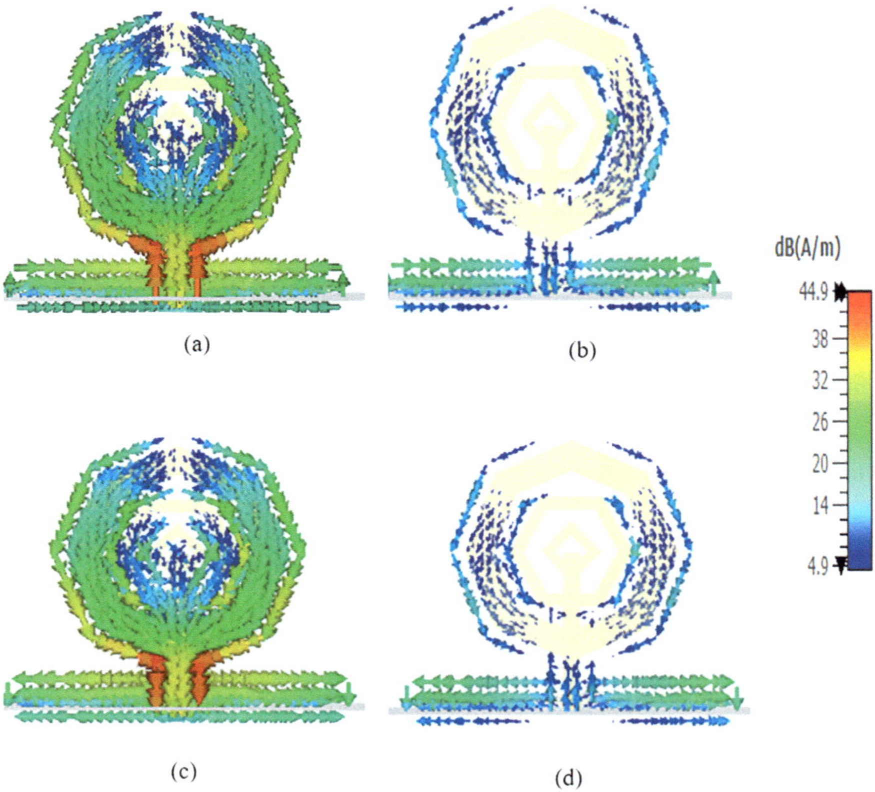

Fig. 6.10 Surface current distribution at different phase angle (**a**) 0°, (**b**) 90°, (**c**) 180°, and (**d**) 270° for proposed antenna design

Antennas play a vital role in modern medical devices by enabling wireless communication for diagnostics, monitoring, and therapeutic applications. In wearable and implantable systems, antennas facilitate real-time data transmission between the human body and external monitoring equipment, enhancing patient mobility and continuous healthcare. They are widely used in devices such as pacemakers, insulin pumps, neurostimulators, and wireless capsule endoscopes. In imaging and diagnostics, antennas support technologies like MRI and microwave imaging, offering non-invasive methods for detecting abnormalities. Additionally, antennas are key components in telemedicine and remote patient monitoring systems, contributing to more efficient and accessible healthcare services. Figure 6.13 shows antenna application in the field of medical for different purposes.

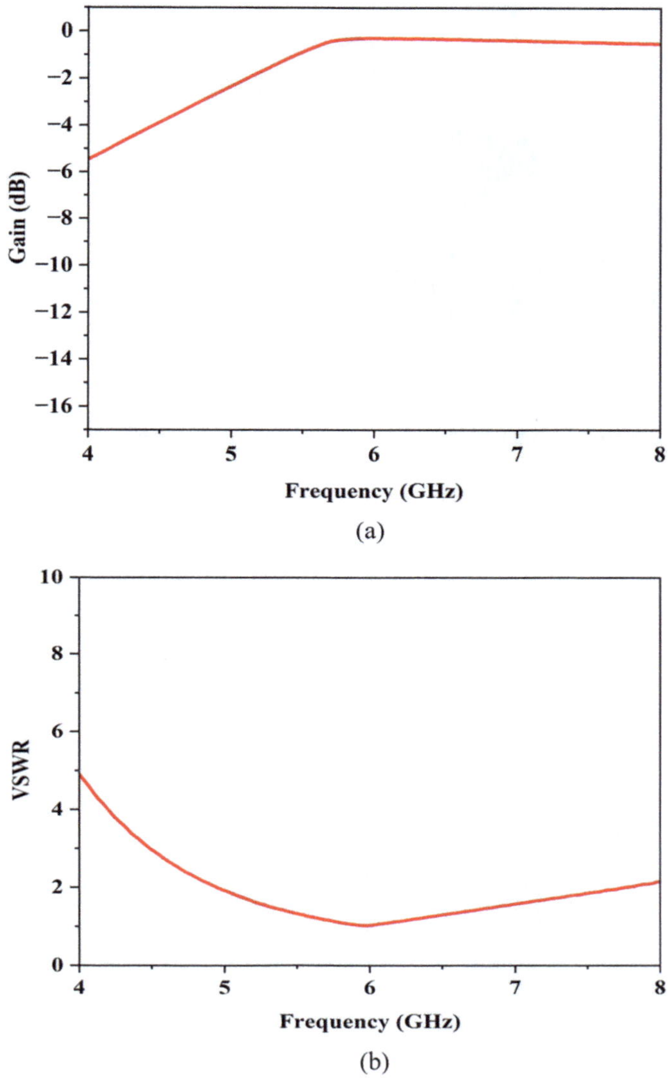

Fig. 6.11 (**a**) Gain and (**b**) VSWR for antenna design

6.7 Open Research Challenges

Implantable medical devices (IMDs) have seen a sharp rise in use in the medical field. IMDs are inefficient, however, due to their implantation and operation issues. This section will outline the implant antenna system's drawbacks and discuss possible research obstacles.

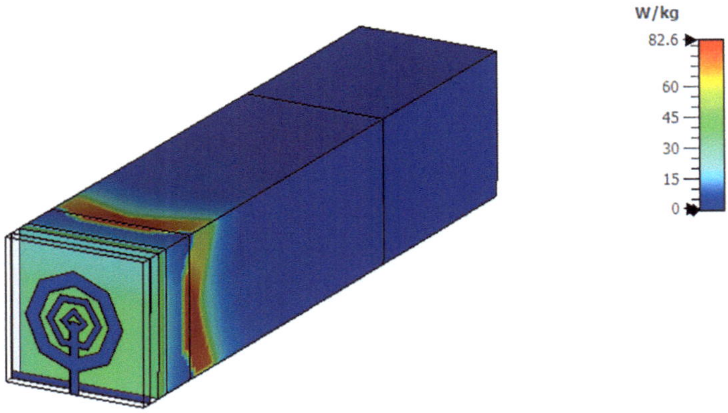

Fig. 6.12 Phantom SAR values for 1 g tissue with 1 watt power for antenna design

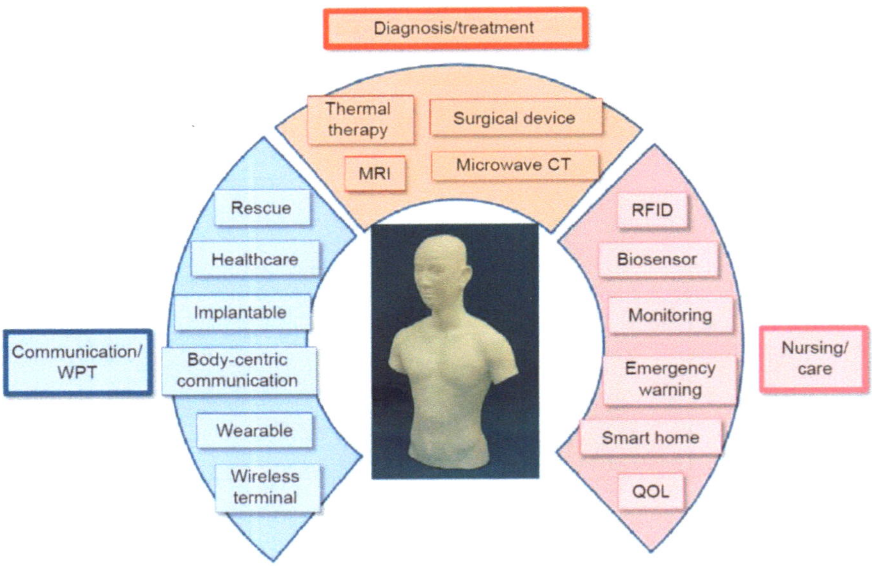

Fig. 6.13 Antenna used in medical devices for different applications [27]

1. An electromagnetic wave is absorbed when coupled with lossy tissue in the far-field and near-reactive regions. A large amount of absorption leads to ineffective antenna operation and low radiation efficiency. To reach the outside receiver antenna, the radiating wave must pass through both the near and far fields because absorption in the far field is inevitable. The biocompatibility of the implant antenna can be addressed close to the field, however, to prevent radio wave absorption. Thus, designing implant antennas that are biocompatible and cover the near field may present a research challenge in the future.

2. The human body is composed of dissimilar biological tissues and organs. Furthermore, the dimensions and properties of biological tissues differ between genders and with time. The detuning effect is therefore regarded as the main research obstacle in the development of an implanted antenna for biotelemetry applications. When placed at a resonance frequency, the implanted antenna might not remain steady for other people or places. The detuning effect was applied to 30 tissues and four anatomical bodies in the research project. There was a noticeable frequency shift of over 70 MHz from the fundamental resonance frequency. The fact that implant antenna design and testing are now restricted to a single tissue environment will be a major drawback for environments with a variety of tissue types. For the future implant antenna to function effectively, it must be studied in many biological settings.
3. When a radiofrequency device is injected into the human body, radiative power absorption may result in long-term, complicated health issues. Thus, the main focus of study will be on efficiently designing implant antennas with a limiting SAR value and choosing biocompatible materials.
4. Using biological energy, a biofuel cell generates electricity without the need for batteries. The main drawback of a biofuel cell is its power at the microwatt level, which restricts its applicability in IMDs. Furthermore, although biofuel cells have been shown to be biocompatible in lab tests, they can destroy tissue cells.
5. There are several battery-free implant antenna power systems available. As an illustration, consider electrostatic generators, piezoelectricity, and thermoelectricity. Even if various power source models have a long lifespan and deliver microwatt-level power, their main research obstacles are biocompatibility and appropriate design.
6. The operational bandwidth is typically limited by traditional antenna dimension reduction approaches. Within the biological milieu, a detuning effect occurs that the narrowband operation cannot avoid. Although biocompatible encapsulation of the implant antenna is used to increase radiation gain and efficiency, it also increases the implant's overall thickness. The design of antennas with sufficient operational bandwidth, radiation efficiency, and gain is therefore still a difficult problem in the field of implantable medical devices [29, 30].
7. Inductive coupling, optical charging, and ultrasonic transducers are the main technologies used in implant antenna systems for wireless power transfer at the milliwatt level. However, the skin temperature increases during the power transmission with those systems, resulting in internal tissue damage and pain. Efficiency can also be decreased by reducing elements associated with power transfer [31, 32].

6.8 Summary

Implantable antenna research has significant potential for enhancing medical monitoring and therapy, making it a fascinating subject of study. Although significant progress has been made in this field, more research is needed to address present challenges and ensure the safety and efficacy of these devices in healthcare settings. In the future, unobtrusive in-body technologies could capture physiologic data from individuals early years. Unobtrusive devices are commonly used by infants, children, and young healthy individuals to monitor heart rate, physical activity, nutritional status, sleep duration, organ function, and breathing rate. Over time, healthcare data can lead to more tailored and cost-effective interventions for medical needs. This big-data method intends to change from reactive and symptom-based medicine to proactive healthcare. Applications include monitoring the elderly, developing countries, military personnel, space exploration, and athletes.

References

1. A. Lamkaddem, A.E. Yousfi, V. González-Posadas, D. Segovia-Vargas, Miniaturized dual band implantable antenna for implanted biomedical devices. IEEE Access **12**, 15026–15036 (2024)
2. S.I.H. Shah, A. Basir, H. Yoo, I.-J. Yoon, A compact ultra-wideband antenna system with stable broadside radiation patterns for brain-machine interface applications. IEEE Trans Antennas Propag (2024)
3. G. Narmadha, M. Malathi, S.A. Kumar, T. Shanmuganantham, S. Deivasigamani, Performance of implantable antenna at ISM band characteristics for biomedical base. ICT Express **8**(2), 198–201 (2022)
4. A. Gupta, A. Kansal, P. Chawla, A survey and classification on applications of antenna in health care domain: Data transmission, diagnosis and treatment. Sādhanā **46**, 68 (2021). https://doi.org/10.1007/s12046-021-01586-4
5. P. Mohanraj, P.R. Selvakumaran, Compact wideband implantable antenna for biomedical applications. Curr Appl Phy **43**, 50–56 (2022)
6. A. Arif, M. Zubair, M. Ali, M.U. Khan, M.Q. Mehmood, A compact, low-profile fractal antenna for wearable on-body WBAN applications. IEEE Antennas Wirel Propag Lett **18**(5), 981–985 (2019)
7. C.-M. Lee, T.-C. Yo, F.-J. Huang, C.-H. Luo, Bandwidth enhancement of planar inverted-F antenna for implantable biotelemetry. Microw Opt Technol Lett **51**(3), 749–752 (2009)
8. G.-P. Gao, C. Yang, B. Hu, R.-F. Zhang, S.-F. Wang, A wearable PIFA with an all-textile metasurface for 5 GHz WBAN applications. IEEE Antennas Wirel Propag Lett **18**(2), 288–292 (2018)
9. K. Aliqab, I. Nadeem, S.R. Khan, A comprehensive review of in-body biomedical antennas: Design, challenges and applications. Micromachines **14**, 1472 (2023). https://doi.org/10.3390/mi14071472
10. Kumar, S. Ashok, and T. Shanmuganantham, Scalp–implantable antenna for biomedical applications. 2020 URSI Regional Conference on Radio Science (URSI-RCRS). IEEE, 2020.
11. S. Ashok Kumar, T. Shanmuganantham, Design of bending antennas for purpose of biomedical applications using novel approach. Trans Electri Electron Mater **22**(6), 725–733 (2021)
12. A.A. Megahed, A.H. Hussein, A.J. Abdullah Al-Gburi, R.H. Elabd, Compact wideband antenna array with DGS-based metamaterial for efficient smartphone communication and SAR reduction. Prog Electromagn Res B **110** (2025)

13. R.H. Elabd, A.J.A. Al-Gburi, Design and optimization of a circular ring-shaped UWB fractal antenna for wireless multi-band applications using particle swarm optimization. Prog Electromagn Res B **106**, 101 (2024)
14. A. Basir, M. Zada, H. Yoo, Compact and flexible wideband antenna for intraoral tongue-drive system for people with disabilities. IEEE Trans. Antennas Propag. **68**(3), 2405–2409 (2020)
15. L.-J. Xu, J.-P. Xu, Z.-J. Chu, S. Liu, X. Zhu, Circularly polarized implantable antenna with improved impedance matching. IEEE Antennas Wirel Propag. Lett. **19**, 876–880 (2020)
16. A. Lamkaddem, A. E. Yousfi, L. E. G. Muñoz and D. Segovia-Vargas, A compact wideband circularly polarized implantable antenna for cardiac pacemaker devices. Proc. Int. Conf. Electromagn. Adv. Appl. (ICEAA), pp. 379–382, 2021.
17. M. Zada, I.A. Shah, H. Yoo, Metamaterial-loaded compact high-gain dual-band circularly polarized implantable antenna system for multiple biomedical applications. IEEE Trans. Antennas Propag. **68**(2), 1140–1144 (2020)
18. M. Yousaf, I.B. Mabrouk, F. Faisal, M. Zada, Z. Bashir, A. Akram, et al., Compacted conformal implantable antenna with multitasking capabilities for ingestible capsule endoscope. IEEE Access **8**, 157617–157627 (2020)
19. A. Lamkaddem, A. El Yousfi, K. A. Abdalmalak, V. G. Posadas, L. E. G. Muñoz and D. Segovia-Vargas, A compact design for dual-band implantable antenna applications. Proc. 15th Eur. Conf. Antennas Propag. (EuCAP), pp. 1–3, 2021.
20. M. Zada, I.A. Shah, J. Nasir, A. Basir, H. Yoo, Empowering remote patient monitoring with a dual-band implantable rectenna system for wireless power and data transfer. IEEE Trans. Antennas Propag. **71**(12), 9509–9522 (2023)
21. A. Basir, M. Zada, Y. Cho, H. Yoo, A dual-circular-polarized endoscopic antenna with wideband characteristics and wireless biotelemetric link characterization. IEEE Trans. Antennas Propag. **68**(10), 6953–6963 (2020)
22. Z. Xia, H. Li, Z. Lee, S. Xiao, W. Shao, X. Ding, A wideband circularly polarized implantable patch antenna for ISM band biomedical applications. IEEE Trans. Antennas Propag. **68**(3), 2399–2404 (2020)
23. G. Samanta, D. Mitra, Miniaturised and radiation efficient implantable antenna using reactive impedance surface for biotelemetry. IET Microw. Antennas Propag. **14**(2), 177–184 (2020)
24. C. Liu, Y. Zhang, X. Liu, Circularly polarized implantable antenna for 915 MHz ISM-band far-field wireless power transmission. IEEE Antennas Wirel Propag. Lett. **17**, 373–376 (2018)
25. M.M. Soliman, M.E. Chowdhury, A. Khandakar, Review on medical implantable antenna technology and imminent research challenges. Sensors **21**(9), 3163 (2021)
26. K.T. Settaluri, H. Lo, R.J. Ram, Thin thermoelectric generator system for body energy harvesting. J. Electron. Mater. **41**, 984–988 (2012)
27. K. Ito, Compelling challenges in antenna technologies for future medical applications. Engineering **11**, 15–17 (2022) ISSN 2095-8099
28. N. Vidal, S. Curto, J.M. Lopez-Villegas, J. Sieiro, F.M. Ramos, Detuning study of implantable antennas inside the human body. Prog. Electromagn. Res. **124**, 265–283 (2012)
29. Y. Qiblawey, M.E.H. Chowdhury, F. Musharavati, E. Zalnezhad, A. Khandakar, M.T. Islam, Instrumented hip implant: A review. IEEE Sens. J. **21**, 7179–7194 (2021)
30. A. Mohan, N. Kumar, Implantable antennas for biomedical applications: A systematic review. Biomed Eng Online **23**(1), 87 (2024)
31. A. Kiourti, K.S. Nikita, A review of in-body biotelemetry devices: Implantables, ingestibles, and injectables. IEEE Trans Biomed Eng **64**(7), 1422–1430 (2017)
32. A. Iqbal, S. Hassan Kiani, M. Al-Hasan, I. Ben Mabrouk, T.A. Denidni, A compact dual-band implantable MIMO antenna for wireless capsule endoscopy. IEEE Transa Antennas Propag **72**, 7515 (2024)

Chapter 7
MIMO (Multiple-Input Multiple-Output) Antennas for Telemedicine

Contents

7.1 Introduction

In recent years, telemedicine has become popular due to advancement of wireless communication technologies. Telemedicine means the treatment of patients from far away distance which makes it easy for patients to access doctors throughout the world [1–3]. In healthcare department, wireless communication technologies help

© The Author(s), under exclusive license to Springer Nature Switzerland AG 2026

A. Segun Ayokunle et al., *Biomedical Antenna Technology*, https://doi.org/10.1007/978-3-031-95268-5_7

telemedicine in many ways. Nowadays, smart healthcare system development is a popular trend, and wireless communication is the pillar of such systems [4]. To serve patients with diagnostics, treatments, and other health services both in hospitals and remotely, every healthcare facility must be equipped with cutting-edge technology. A rapid response unit is also necessary to address the surge of patients and questions produced during disasters [5]. The existing healthcare system is primarily reliant on fourth generation (4G) networks to allow smart healthcare applications; but, due to limited bandwidth resources, these networks cannot sustain the fast expansion of these applications in the long term. Compared to 4G, fifth generation (5G) networks offer superior services in terms of bandwidth, dependability, latency, and data rate [6]. Figure 7.1 shows application of 5G technology in different purposes in health sector.

In the revolution of cellular generation, single-input single-output (SISO) technology is not capable. As the use of the internet increases drastically in various areas, the wireless system is in demand if it offers high data rates and satisfactory channel capacity [7]. To fulfill the demand of the internet, a technology called multiple-input multiple-output (MIMO) comes into the picture. MIMO is an antenna technology in which a wireless communication system uses multiple antennas at the transmitter and the receiver [6]. The antennas at each end of the communication circuit are combined to minimize the bit errors, optimize data speed, and improve the capacity of radio transmissions by enabling data to travel over many signal paths at the same time [8]. A pictorial representation of MIMO technology with multiple antennas is shown in Fig. 7.2.

This chapter provides a MIMO antenna design and its analysis for various healthcare applications. In the area of MIMO antennas, several research studies and efforts have been conducted to reduce mutual coupling and propose many approaches in

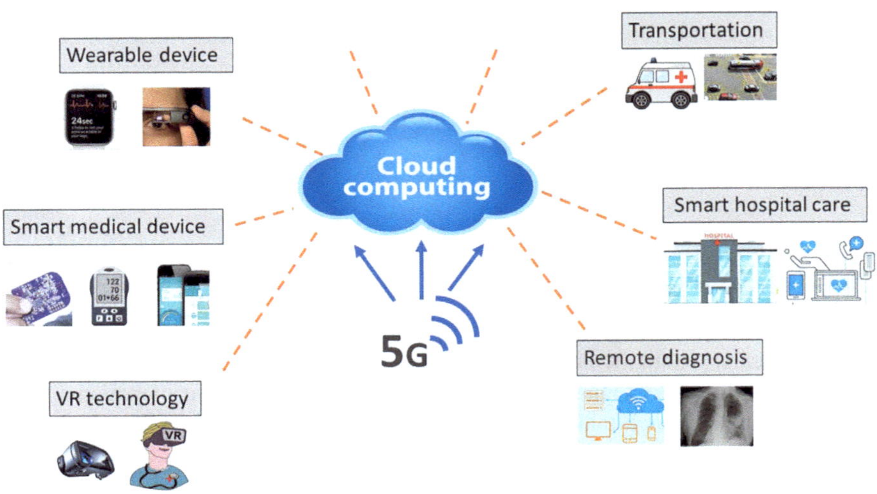

Fig. 7.1 Showing applications of 5G technology in healthcare [6]

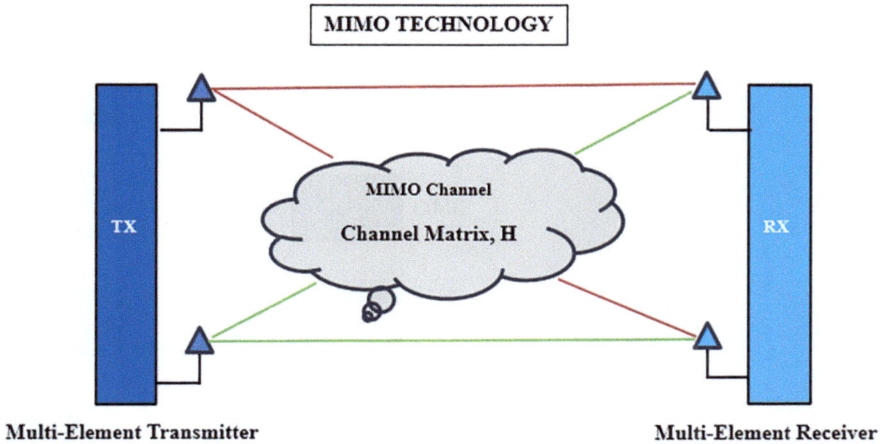

Fig. 7.2 MIMO technology with multiple antennas on the transmitter and receiver sides

order to achieve the MIMO antenna array performance. Frequency range: Application-specific (e.g., 2.4 GHz or 5 GHz). For wireless communication, MIMO is an improved version of microstrip antenna technology with multiple antennas [5]. It is the foundation of current antenna technology and the wireless generation of the future. Multiple antennas are used in both the MIMO transmitter and receiver to enhance wireless communication performance.

Spatial multiplexing makes use of multipath propagation. By utilizing Shannon's capacity theorem, MIMO technology aims to enhance the demands of wireless communication, including high data rates, low latency, increasing channel capacity, improving quality and spectral efficiency, decreasing multiple fading, low cost, and compact size. MIMO must adhere to high SNR requirements. For high data speeds and optimal signal quality, MIMO devices are utilized to send and receive data more quickly than a single antenna can do so simultaneously. The development of MIMO technology has allowed Richard Roy, Bjorn Ottersten, Arogyaswami Paulraj, Thomas Kailath, Greg Rayleigh, Gerard J. Foschini, and Emre Telatar to play a significant role in enhancing channel characteristics, increasing data rate, and improving channel connectivity [9]. Figure 7.3 shows various configurations of antenna systems.

There are various variations of MIMO technology including massive MIMO, SU-MIMO, and MU-MIMO, and MIMO offers many benefits compared to SISO, SIMO, and MISO technologies. In the first figure, a single antenna appears to be used by both the transmitter and the receiver in a single-input single-output (SISO) system. The second one, known as receiver diversity, uses single-input and multiple-output (SIMO) technology to demonstrate that the transmitter has a single antenna and the receiver has two. Multiple-input and single-output (MISO) refers to the situation when a transmitter has two antennas and a receiver has one; this is also known as transmitter diversity. MU-MIMO addresses the drawbacks of SU-MIMO. In both

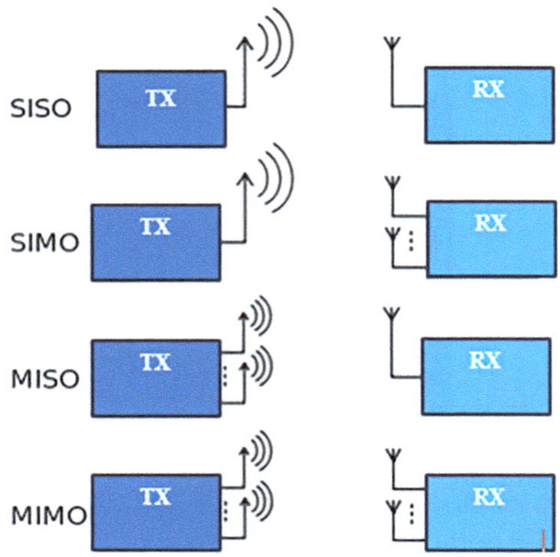

Fig. 7.3 Various configuration of antenna systems [7]

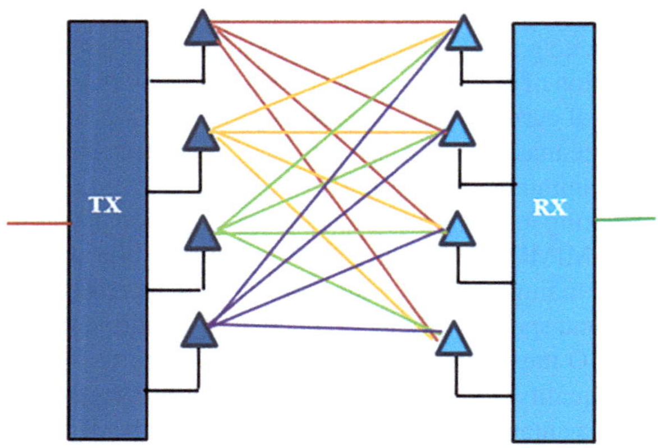

Fig. 7.4 Pictorial representation of MIMO system

MU-MIMO and massive MIMO, the transmitter and receiver have used many antennas. Pictorial representation of MIMO system is shown in Fig. 7.4.

Information can be broadcasted or communicated at the speed of light with MIMO. It is a key component of 5G, 6G networks and is utilized in 3G, Wi-Fi, 4G, LTE, WLAN, and WiMAX technologies [10, 11]. MIMO technology offers improved separation between high-gain elements and operates in many bands. MIMO makes signals stronger because MIMO provides higher throughput, video, and other large-scale material that may travel over a network more quickly and in enormous quantities. Several data streams use MIMO to reduce the likelihood of lost data packets while enhancing visual and aural quality. High-bandwidth communications use MIMO. As illustrated in Fig. 7.4, MIMO has three primary purposes for handling signal transmission and reception over multipath channels (where Tx represents multiple transmitting antennas, Rx represents multiple receiving antennas, and H is the channel).

MIMO enhances wireless communication through three primary functions such as precoding, diversity coding, and spatial multiplexing in meeting user coverage and data throughput requirements [3]. For high gain, precoding is applied at the transmitter and improves beam shaping. Bell Laboratory introduces spatial multiplexing to enhance the MIMO system's performance metrics. The benefit of precoding is combined with numerous streams, and both the transmitter and the receiver need to be aware of channel state information (CSI). Without the CSI, spatial multiplexing produces improved channel capacity increase. Precoding in conjunction with spatial multiplexing results in increased capacity and high gain at the receiver. Signal diversity is improved by diversity coding. Combining multiplexing and diversity coding improves wireless channel performance [12].

7.2 MIMO Antenna Design for Telemedicine Applications

CST Microwave Studio has been used to create the single monopole microstrip antenna using split ring resonator (SRR). Figure 7.5a and b show the geometry of the single element with partial ground ring and SRR. A partial ground ring and CSRR are used in the design of the single monopole microstrip element. The structures are 20 mm × 20mm × 1.6 mm.

Figure 7.6a shows the prospective view of single-port antenna. Figure 7.6b shows SRR structure. In order to attain resonance at a certain frequency range, SRR must

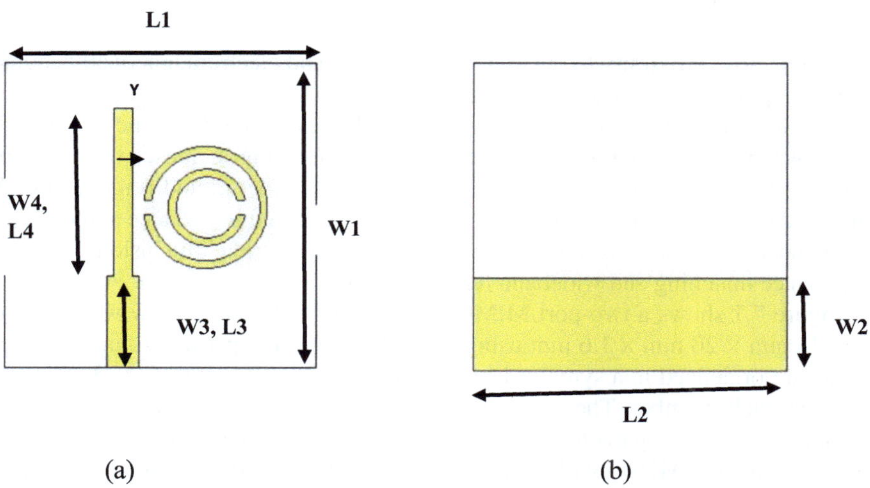

Fig. 7.5 A single-port antenna design: (**a**) top view and (**b**) back view

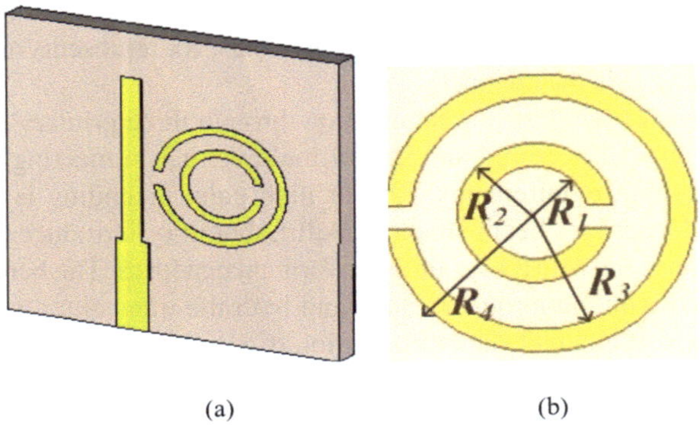

Fig. 7.6 (**a**) Prospective view single-port antenna. (**b**) Split ring resonator

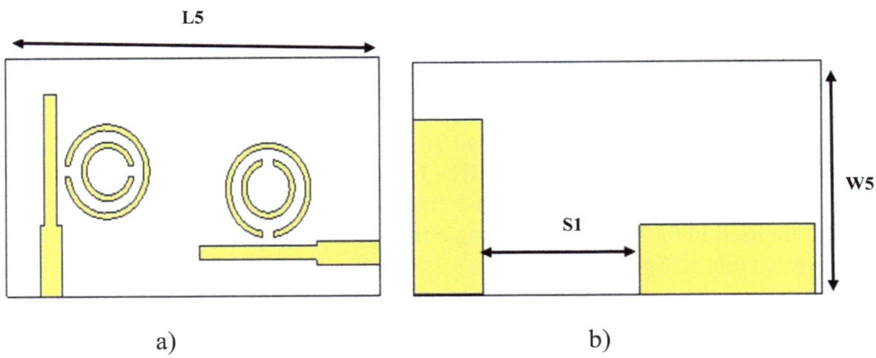

Fig. 7.7 A two-port MIMO antenna (**a**) top view and (**b**) back view

remain in close proximity to the monopole radiator and electromagnetically couple to a monopole antenna. The coupling capacitance decreased as the distance between the feedline and SRR increased, which caused the resonating frequency to change to a lower value. By acting as a reflector, the ground ring increases the resonant isolation by reducing the occupied coupling currents. The ground ring has a high current density, and it has become the part of the improved impedance and enhancing bandwidth area. Any change in SRR and ground ring dimensions impacts impedance matching and wideband resonance behavior.

Figure 7.7 shows a two-port MIMO antenna top and bottom view with dimension 35 mm × 20 mm × 1.6 mm using SRR. The SRR and partial ground ring elements are organized in a symmetrical rotating configuration, with a 90-degree gap between each member. The resonance range swings to a lower frequency as the distance between the parts increases. To improve impedance matching at frequency 5GHz, monopole MIMO with SRR and ground ring is used. It is suitable for telemedicine applications. The elements with CSRR and partial ground ring are arranged rotational and symmetrical structure with 90 degrees interval of each.

When increasing the distance between the elements, the resonance range shifts to a lower frequency. Monopole MIMO with SRR and ground ring make better impedance matching at resonant frequency of 5 GHz [8].

7.2.1 Frequency Range: Application-Specific (E.G., 2.4 GHz or 5 GHz)

MIMO antenna design configurations can be 2 × 2, 4 × 4, and 8 × 8. Compared to SISO antenna, MIMO antenna of 2 × 2, 4 × 4, and 8 × 8 increases the performance roughly 30% to 90% in connection speeds. MIMO antennas transmit large amounts of data across the same spectrum. Modern devices use 4 × 4, 8 × 8, and 2 × 1 for mid- and high-frequency cellular bands, and 2 × 1 or 2 × 2 for low-frequency bands. However, the performance metrics of the various MIMO setups vary. MIMO antenna works in different diversity modes [9]. Time diversity: Information can be sent at different intervals of time. Frequency diversity: Diversity uses different frequencies, and it is in the form of different channels or technologies. Space diversity: Space diversity in its broadest sense and used antennas in many locations to take advantage of the multipath propagation in terrestrial environments.

MIMO is utilized in various applications, and a few of the services include Bluetooth operating at 2.480GHz, WiMAX at 3.3–3.7GHz, Wi-Fi operating at 2.4–5GHz, WLAN at 5.1–5.825GHz, Ka-band at 33.4–36.0GHz, UWB at 3.1–10.6GHz, and 5G NR N77 at 3.3–4.2GHz, N79 at 4.4–5.0GHz, N78 at 3.3–3.8GHz [13]. Moreover, industrial mm-wave radar sensors operate at 57–64GHz, IEEE 802.11ad at 57–71GHz, and high-speed authorized microwave data links at 36–40GHz. The design of multifunctional patterns with polarization reconfigurable antennas and 5G small cell base station antennas presents sophisticated challenges for future MIMO system design. These challenges include maintaining low correlation, ensuring compact size and planar structure, and achieving good isolation and high gain over a frequency range for wideband, ultra-wideband, narrowband, and dual-band applications. This paper extensively discusses these challenges and their associated designs [14].

7.2.2 Corporate Feed

A corporate feed in antenna design is a network used to distribute power equally to multiple radiating elements, ensuring uniform phase and amplitude across the array. Typically implemented using microstrip lines, waveguides, or coaxial cables, the corporate feed is characterized by its branching structure, where power is divided equally at each junction to maintain consistent signal strength and performance. This design is crucial for applications requiring high gain directional antennas, such as radar systems, satellite communications, and phased arrays, as it minimizes mismatches and maximizes efficiency. Proper design of the corporate feed ensures reduced losses and optimal electromagnetic performance across the operating frequency range.

7.2.3 *Hybrid Feed*

A hybrid feed in antenna design combines features of both corporate and series feeding techniques to balance power distribution, phase uniformity, and design complexity. In this approach, the feed network integrates the corporate feed's power-splitting capability with the series feed's simplicity and compactness. Hybrid feeds are often used in large antenna arrays where achieving efficient power distribution and maintaining a low profile are critical. This method optimizes antenna performance by reducing losses, simplifying the feed network layout, and achieving a compromise between design complexity and cost. Hybrid feeds are commonly applied in modern phased arrays, wireless communication systems, and radar applications.

7.2.4 *Side Lobe Levels Reduction by Taylor Amplitude Weighing/Tapering Technique*

The Taylor amplitude weighting or tapering technique is a widely used method in antenna design to reduce side lobe levels, improving radiation pattern performance and minimizing interference. This technique involves adjusting the amplitude of excitation across the antenna array elements based on a mathematical distribution derived from the desired side lobe level and beamwidth trade-offs. By applying this weighting, the main lobe remains sharp and focused while side lobes are suppressed, enhancing the antenna's directional properties and reducing unwanted signal pickup. The Taylor method is particularly effective in radar, satellite communications, and other applications where low side lobe levels are critical for minimizing noise and enhancing signal clarity.

7.3 Placement Configuration of Frequency-Selective Surfaces for Antenna's Gain Enhancement

This is an effective technique to improve antenna gain. In this section, FSS unit cell has been proposed to enhance the gain of the antenna. The design and analysis of the proposed antenna are explained in this section. Repeating unit cells make up the planar structures known as frequency-selective surfaces (FSSs). It is possible to configure the unit cells in one or two dimensions. The size, arrangement, and spacing of the unit cells define the resonance frequency of the FSS. FSSs can emit, receive, or reflect electromagnetic spectrum energy depending on the dielectric substrate, polarization, angles of frequency, and resonant frequency. FSSs generally fall into two categories: patch and slot. Patch FSSs serve as low-pass filters, whereas slot FSSs serve as high-pass filters.

The suggested metallic hexagon FSS unit cell (20 mm × 20 mm × 1.6 mm) is shown in Fig. 7.8. The thickness of FR4 substrate is 1.6 mm with a dielectric constant of 4.3 is used to install the metallic hexagonal ring FSS by using CST Microwave Studio executes the intended FSS. Figure 7.9 shows the S-parameters for the FSS unit cell structure. Research-grade hexagonal ring FSSs have a broad band-stop frequency range from 4 to 6 GHz, resonant at 5GHz, and are more significant. The FSS exhibits band-stop behavior with a bandwidth of 2 GHz and a deepest level of the $|S21| < -30$ dB, as shown by the S-parameter. In both transverse magnetic (TM) and transverse electric (TE) modes, the S-parameter values are the same. Because of its angular stability, the proposed FSS unit cell is further employed in MIMO antennas for gain augmentation and MC reduction.

Metallic FSS Array Reflector: An FSS can be used to improve the gain, the bandwidth, and the isolation of the antennas. An FSS is a recurring configuration of tiny structures that has the ability to regulate electromagnetic wave transmission and reflection [15]. The intended unit cell is intended to run between 4 and 6 GHz. It is printed on FR4 substrate that is 1.6 mm thickness by using CST and simulates the unit cell. The ultimate optimal unit cell design is a hexagonal geometric pattern, as seen in Figure. Figure displays the port assignments and boundary conditions used to stimulate the unit cell. Figure shows the reflection coefficient curve of the unit cell, which indicates that it is resonant at 5 GHz (2–4 GHz). Furthermore, the transmission coefficient plot reveals that the FSS unit cell exhibits almost no loss at the resonant frequency. Thus, almost all the electromagnetic waves contained in the intended as shown in Figure, frequency bands are communicated through the unit cell [16] (Fig. 7.10).

Figure 7.11 shows different views of two-port MIMO antenna with FSS. An FSS surface needs to be able to cover the same bandwidth as the recommended MIMO antenna in order to be designed to satisfy antenna applications' requirements. Both

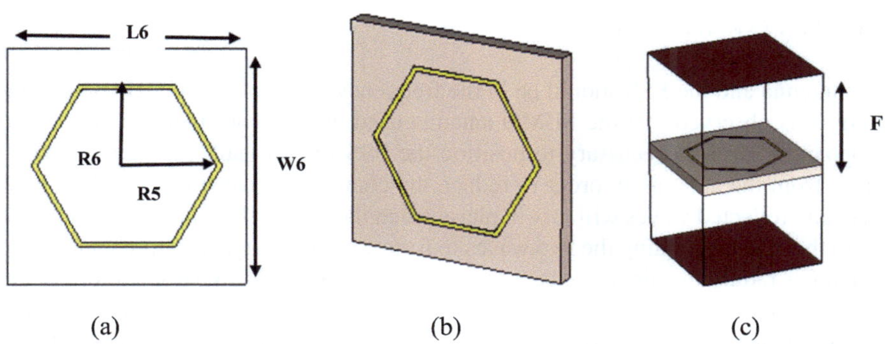

Fig. 7.8 Metallic hexagonal shape of frequency-selective surface

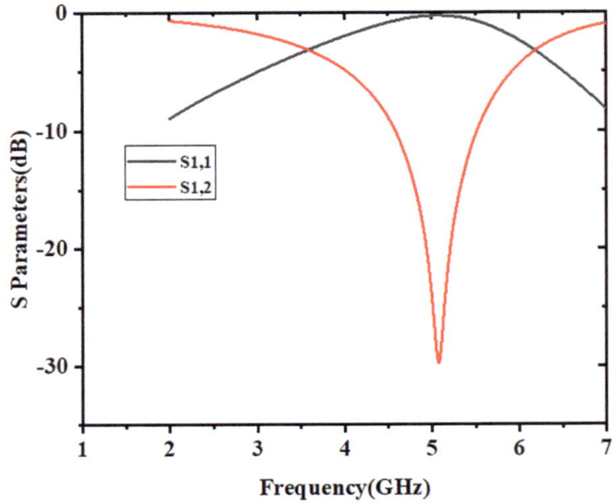

Fig. 7.9 Return loss vs frequency for unit cell structure

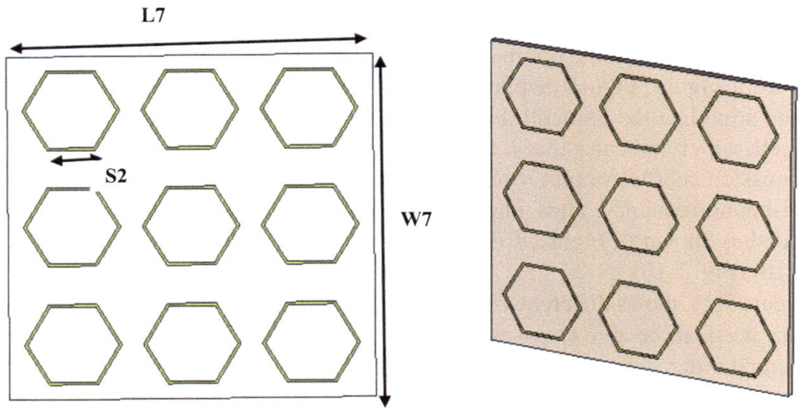

Fig. 7.10 FSS array structure

the antenna and the FSS should be in the frequency range of 4 to 6 GHz when running. It is observed that the MIMO antenna used in this work is integrated with a stopband FSS. It is necessary to position the FSS surface half a wavelength ($\lambda/2$) apart from the antenna in order to reduce the coupling between the two. The incident and reflected waves will mix in phase when the FSS surface is positioned at $\lambda/2$ from the antenna, adding the two waves to form a stronger signal. This enhances the antenna's radiation gain, or its capacity to focus its energy on a particular direction.

The distance between the FSS layer and the MIMO antenna is specified in the proposed work as h_{FSS} = 10.5 mm. In addition, 3 × 3 unit cells are periodically arranged in the space of S2 to make an FSS array. $W_{FSS} \times L_{FSS} \times T$

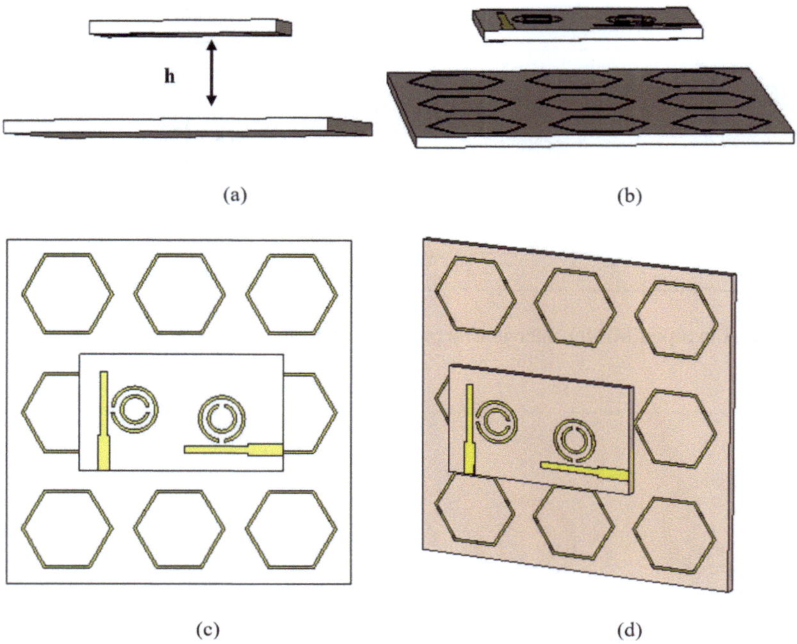

Fig. 7.11 A two-port MIMO antenna with slots (**a**) antenna and FSS separation side view, (**b**) antenna and FSS separation perspective view, (**c**) antenna with FSS top view, (**d**) antenna with FSS perspective view

(60 mm × 60mm × 1.6 mm) are the dimensions of the FSS array layer. The FSS array is placed behind the MIMO antenna at a distance of h_{FSS} in order to enhance antenna performance, as indicated in Figure. The inclusion of FSS (Full Cells of FSS array) only slightly enhances the MIMO antenna's performance, as seen in Figure. The suggested MIMO antenna's bandwidth has an impact on the reflection coefficient which decreases in frequency and increases in maximum gain from 2.5 dBi to 4 dBi. This situation calls for an innovative method when h_{FSS} change increase or decrease, gain will change as like in this case, a new method for increasing the MIMO gain without sacrificing the MIMO antenna's other performance attributes is needed.

7.4 Planar Microstrip Array Antenna Bandwidth Enhancement Techniques

7.4.1 Parasitic Patch Microstrip Antenna

In MIMO antennas, the parasitic element is employed to increase bandwidth and isolation, effectively mitigating the impact of correlation between ports. By strategically positioning the parasitic element, mutual coupling within the operating

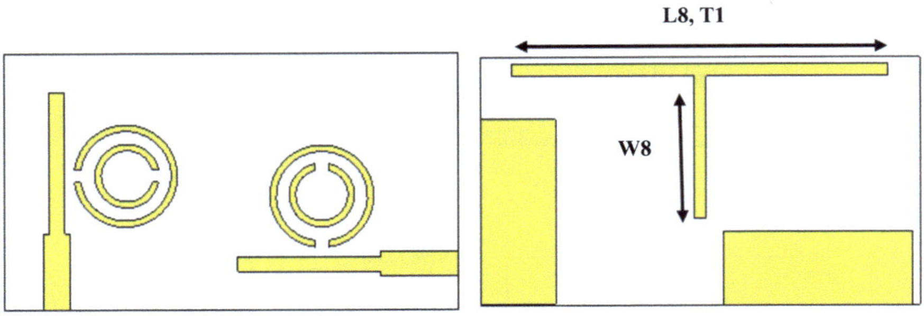

Fig. 7.12 A two-port MIMO antenna with parasitic patch

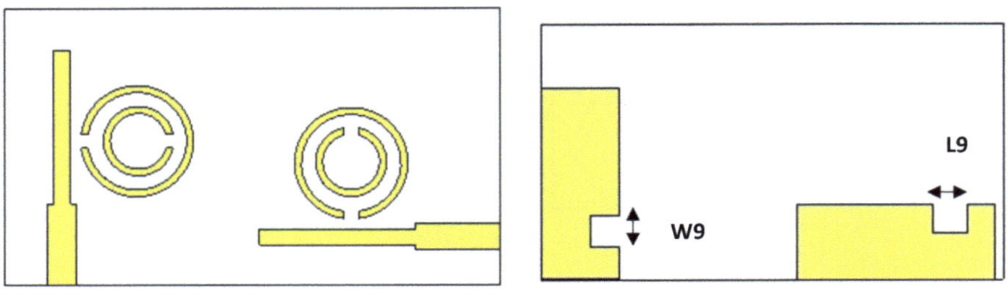

Fig. 7.13 A two-port MIMO antenna with slots

frequency band is eliminated. SRR and CSRR concepts are implemented in MIMO to minimize mutual coupling and improve isolation among radiators. Utilizing SRR and CSRR in MIMO antennas enables wideband and dual-band operation, while also enhancing isolation between ports. Metamaterial and FSS are vital techniques in MIMO to reduce antenna size and boost isolation with high gain. The incorporation of a decoupling structure enhances the performance metrics of MIMO and aids in generating superior far-field patterns, as well as achieving impedance matching and reducing mutual coupling between elements. A T-shape parasitic patch element is loaded in the two-port MIMO antenna to extended bandwidth but it reduces resonance range (Fig. 7.12).

7.4.2 *Using Slots at the Microstrip Antenna Radiators*

Loading a U-shaped slot in the two-port MIMO can increase bandwidth but may also cause a shift in resonance range to lower values and slight decrease in gain and directivity (Fig. 7.13).

7.4.3 *Proximity-Coupled Feed Microstrip Antenna*

A proximity-coupled feed microstrip antenna is a configuration where the radiating patch is excited through electromagnetic coupling via a feed line separated by a dielectric substrate. Unlike direct-feed techniques, the feed line does not make physical contact with the patch, improving impedance matching and reducing surface wave losses. This method offers increased bandwidth and flexibility in design, making it suitable for high-frequency applications. However, it introduces design complexities, such as precise alignment of the feed line and the patch and the need for multilayer fabrication. Proximity coupling is particularly advantageous in modern wireless communication systems, where enhanced performance and reduced interference are essential.

7.4.4 *Log-Periodic Microstrip Antenna*

A log-periodic microstrip antenna is a type of frequency-independent antenna designed using the log-periodic principle, where the geometry of the antenna elements is scaled logarithmically to achieve a wide operational bandwidth. The radiating elements are typically arranged in a planar configuration on a microstrip substrate, enabling compact and lightweight designs. This antenna exhibits stable radiation characteristics and impedance across a broad frequency range, making it ideal for applications such as broadband communication, electromagnetic spectrum monitoring, and radar systems. Challenges in designing log-periodic microstrip antennas include maintaining consistent performance over the desired frequency range and addressing fabrication complexities due to the intricate scaling of elements.

7.5 Problem Formulation

With the increasing demand for remote healthcare services (telemedicine), there is a need for reliable, high-speed, and interference-free communication between healthcare providers and patients. MIMO (Multiple-Input Multiple-Output) antenna systems play a crucial role in enhancing the performance of wireless communication by improving data rates, reliability, and coverage. The challenge is to design a MIMO antenna that meets the strict requirements of telemedicine applications, including low latency, high data throughput, and robust connectivity. In view of the available literature and applications, the design optimizes a MIMO antenna system for telemedicine applications that provide high data rates, low latency, and reliable connectivity while ensuring compact size and low-power consumption is the problem definition.

7.6 Related Work

Improving 5G communication requirements such as high data speeds, low latency, improved channel capacity, higher quality, greater spectrum efficiency, decreased multiple fading, low cost, and compact size is the primary goal of MIMO technology [11]. In order to achieve a small size with several antenna components on the substrate, mutual coupling (MC) must be managed on the space length of the two elements [16]. This presents a major design difficulty for MIMO structures. Several approaches have been proposed in recent years to address the issue of high mutual coupling in MIMO antennas. Several design strategies, such as complementary split ring resonators (CSRR), defective ground structures (DGS), decoupling components, parasitic elements, neutralization lines and slots, line resonators, isolation technique, EBG (electromagnetic band gap), metamaterial, and metasurface are utilized to provide low mutual coupling between antennas [16]. An extensive thorough investigation of diversity and mutual coupling reduction techniques in compact MIMO antennas has been carried out in [20].

7.7 Simulation Results and Discussion

The input–output interactions between ports in an electrical system are explained by using scattering parameters. The single-port antenna and four-port monopole MIMO antenna simulated S characteristics (reflection and transmission coefficient) are shown in Fig. 7.14. The four-port MIMO antenna impedance bandwidth and the single-port element bandwidth are from 4GHz to 6.4GHz. Figure 7.15 shows the simulated S-parameters such as S11, S12, S13, and S14 with mutual coupling between the ports. To ensure that the simulated isolation between the elements is greater than 10 dB across the entire dual band. The MIMO configuration having design-1 provides dual-band, whereas design-2 provides wideband response within the working band. The distance between two elements determines their mutual connection. As the elements between farther apart, their isolation gets better, and their impedance rises as well.

Radiation pattern: The polarization and magnitude of the field emitted by a transmitting antenna as a function of the angle relative to the antenna are explained by the radiation pattern of the monopole MIMO antenna. Figure 7.16a, b and c show the radiation pattern plot for MIMO antenna, MIMO antenna with parasitic element, and MIMO antennas with FSS, respectively. Illustrates the radiation pattern of the presented MIMO. For the two primary E and H planes, the far-field radiation pattern was simulated at specific frequencies,such as 5GHz, in the operational

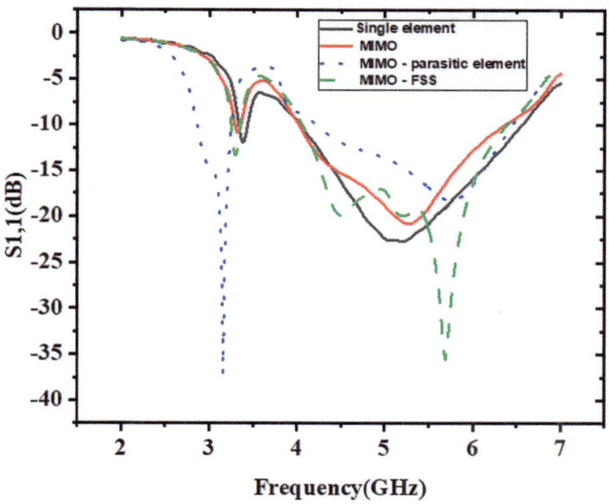

Fig. 7.14 Return loss (S11) versus frequency plot for different antenna configurations

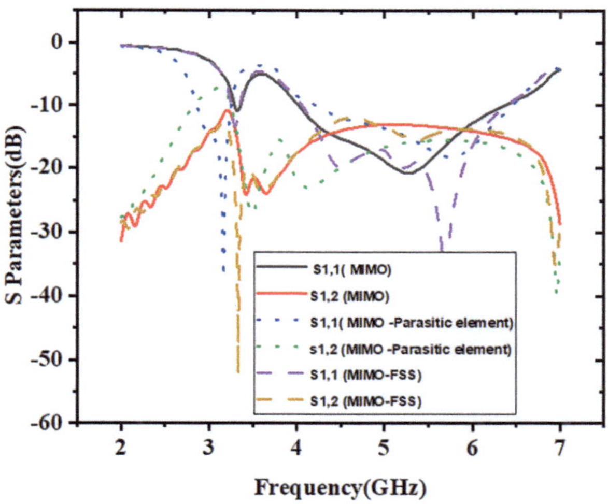

Fig. 7.15 S-parameters (dB) versus frequency (GHz) plot

bands. The simulation output is to activate port 1 and close the remaining ports with matching 50-ohm loads. MIMO antenna directional radiation patterns are more stable in the suggested band, depending on the antenna position. It can receive signals from all directions and emit omnidirectional radiation.

Directivity and Gain: The concentration of an antenna radiation pattern in a given direction is measured by directivity, while gain shows the concentration of input power in a particular direction. Figure 7.17 shows the gain of the MIMO antenna, and Fig. 7.18 shows the directivity of the proposed MIMO antenna, MIMO

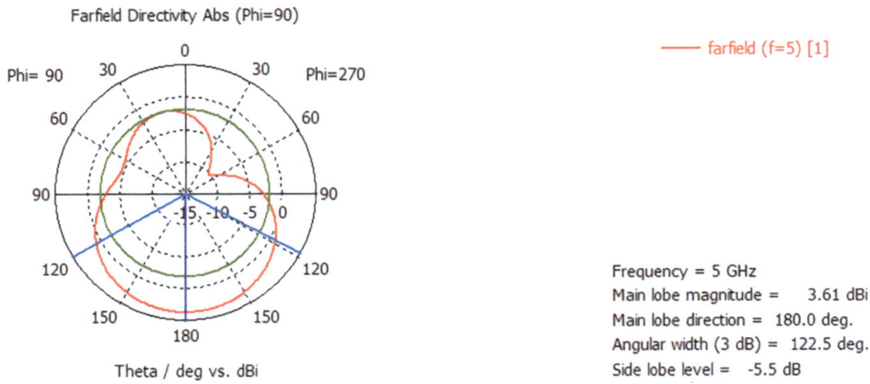

(a)

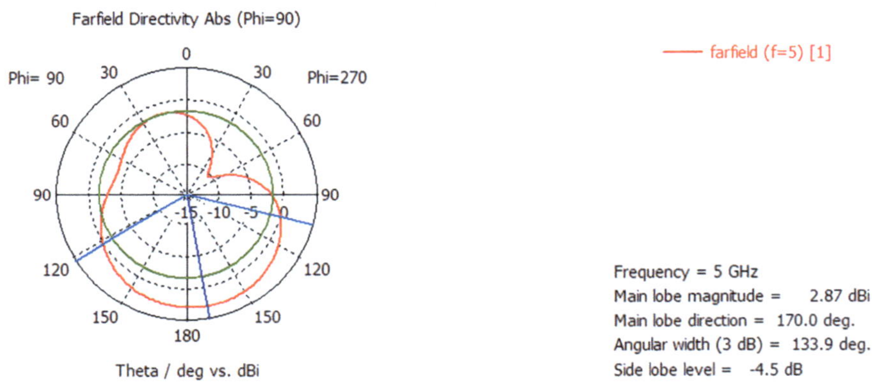

(b)

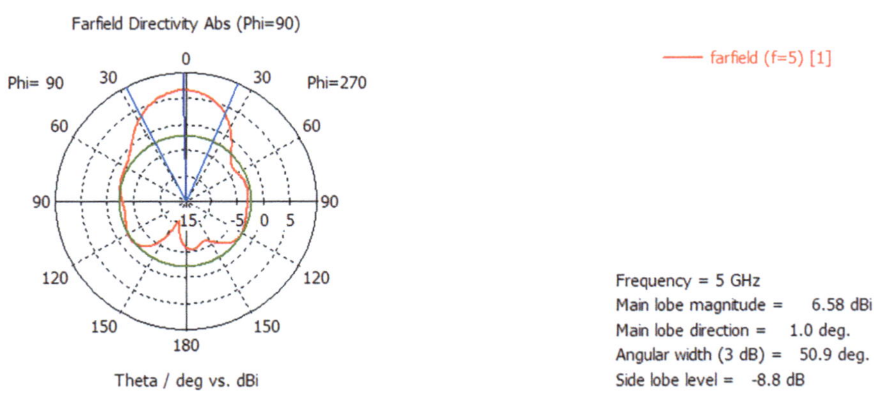

(c)

Fig. 7.16 Far-field directivity plots for (**a**) MIMO antenna, (**b**) MIMO antenna with parasitic element, (**c**) MIMO antenna with FSS

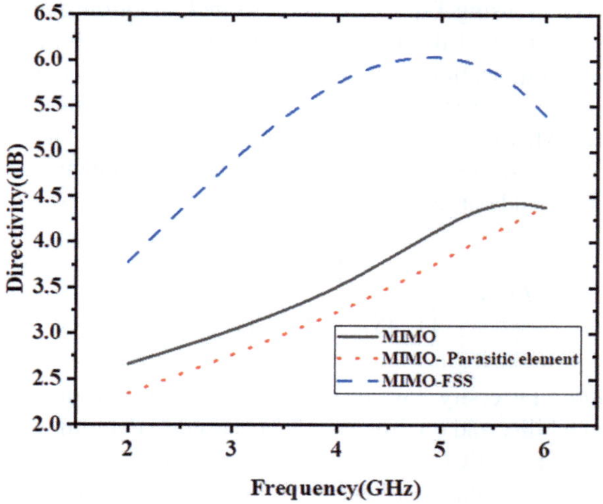

Fig. 7.17 Directivity versus frequency plot for MIMO, MIMO with parasitic element, and MIMO with FSS

Fig. 7.18 Gain in dB versus frequency plot for MIMO, MIMO with parasitic element, and MIMO with FSS

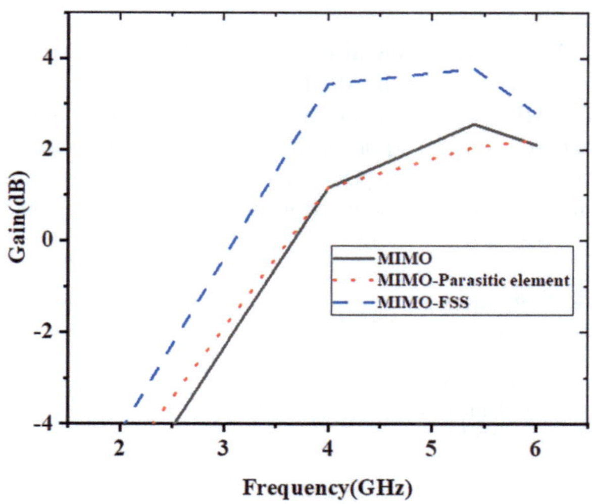

with parasitic element, and MIMO with FSS. It is observed that the simulated value of gain and directivity of a monopole MIMO antenna is 4 dB within the different operating frequencies. When increasing isolation between the elements, gain and directivity value is increased as shown in Fig. 7.6 of the radiation pattern.

Envelope Correlation Coefficient: Another diversity parameter such as envelope correlation coefficient (ECC) defines MIMO configurations correlation. By applying the S-parameter method, the ECC of the MIMO antenna can be evaluated (where i^{th} & j^{th} is the port of the elements and S^{ij} is the index element of the scattering matrix) by using Eq. (7.2). The channel situation makes it difficult to establish good variety, and the ECC value is high. For the best diversity performance MIMO communication system, a low ECC value is accepted. It has a value of 0 or 1 (1 indicates a strong correlation and zero indicates no correlation at all). ECC for MIMO, MIMO with parasitic element, and MIMO with FSS has been shown in Fig. 7.19 and the value of ECC in this MIMO antenna is < 0.01, with a lower value of ECC indicating better pattern variety.

$$ECG_s = \frac{\left|S_{ii}^* S_{ij} + S_{ji}^* S_{jj}\right|^2}{\left(1-\left|S_{ii}\right|^2-\left|S_{ji}\right|^2\right)\left(1-\left|S_{jj}\right|^2-\left|S_{ij}\right|^2\right)} \tag{7.1}$$

Diversity Gain: Diversity gain in a wireless system is a measure of a MIMO antenna's dependability and quality. Eq. (7.2) is used to determine the DG and substituted the value of the ECC < 0.01 in it. The diversity gain value in this MIMO antenna is > 9.95 dB within the allowable frequency band indicating good diversity performances. Diversity gain versus frequency plot has been shown in Fig. 7.20 for MIMO, MIMO with parasitic element, and MIMO with FSS.

$$DG = 10\sqrt{1-(ECC)^2} \tag{7.2}$$

Mean Effective Gain: Mean Effective Gain (MEG) is defined as the ratio of the power received by the MIMO antenna to the power received by an isotropic antenna. The MEG can be calculated in terms of S-parameters by using eqs. (7.3) and (7.4) (where i & j is the port of the element). MIMO antenna port has equal power level to get best diversity performance, and the MEGi values should be between -3 dB and -12 dB throughout whole frequency range as shown in Fig. 7.21. The power

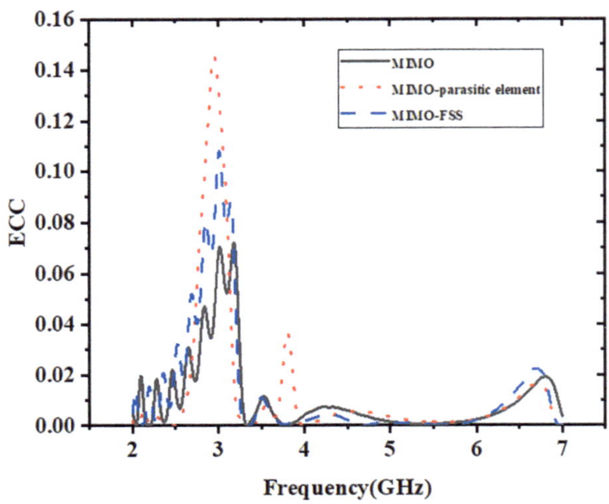

Fig. 7.19 Envelope correlation coefficient versus frequency plot for MIMO, MIMO with parasitic element, and MIMO with FSS

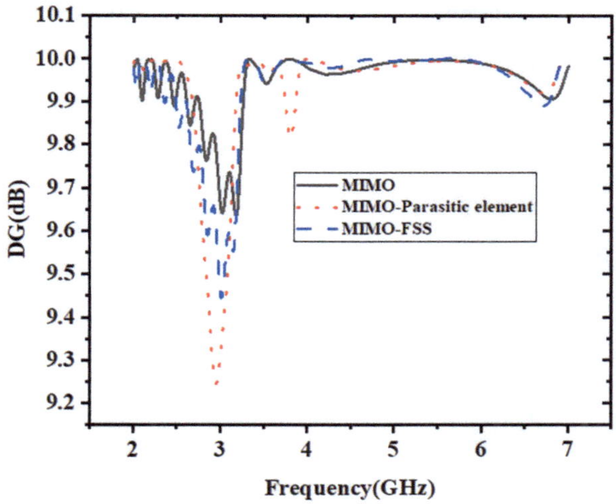

Fig. 7.20 Diversity gain in dB versus frequency in GHz plot for MIMO, MIMO with parasitic element, and MIMO with FSS

relation between any two ports should not be greater than 3 dB. To view results, confirm of the designed MIMO antenna have good diversity performance.

$$MEG_i = 0.5\left(1 - |S_{ii}|^2 - |S_{ji}|^2\right)$$

$$MEG_j = 0.5\left(1 - |S_{ij}|^2 - |S_{jj}|^2\right)$$

$$\left(MEG_i - MEG_j\right) < 3dB$$

$$\frac{MEG_i}{MEG_j} = \pm 3dB \tag{7.3}$$

Total Active Reflection Coefficient: The Total Active Reflection Coefficient (TARC) is the ratio of the total reflected power from the radiating elements to the total incident power on the patch. TARC value for the MIMO antenna should be less than 0 dB to show the best performance of the MIMO antenna throughout whole operating frequency band. The relation between TARC and S-parameter is described in Eq. (7.4). The value of TARC is low to show good performance of MIMO system and better data quality with higher transfer rates.

$$TARC = \sqrt{\frac{\left(|S_{ii} + S_{ij}e^{j}\theta|^2\right) + \left(|S_{ji} + S_{jj}e^{j}\theta|^2\right)}{N}} \tag{7.4}$$

where i and j are the port numbers, θ denotes the phase angle between two ports, and N is the number of radiating elements [17] (Fig. 7.22).

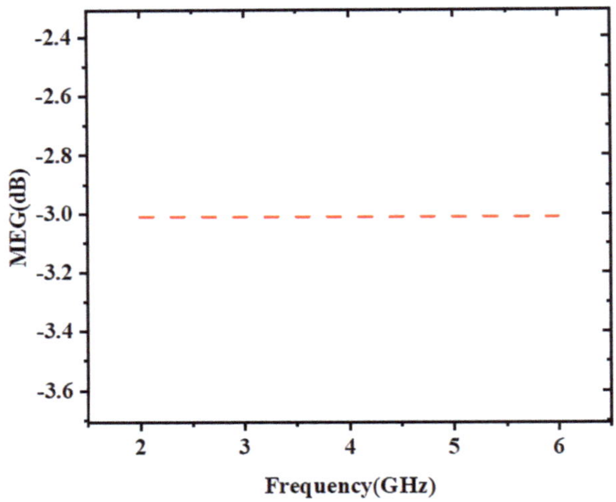

Fig. 7.21 MEG in dB versus frequency plot

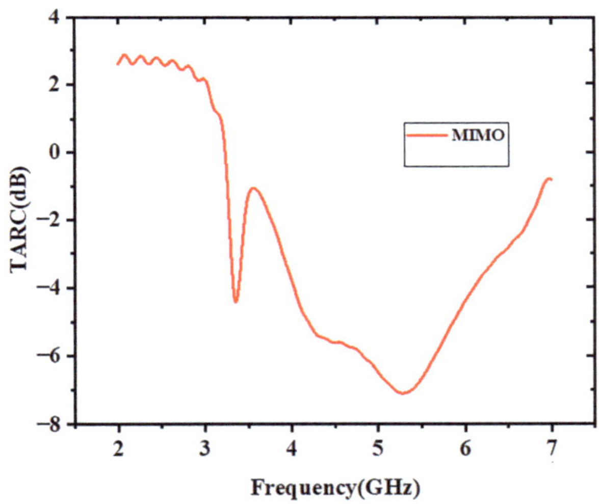

Fig. 7.22 TARC (dB) versus frequency (GHz) plot

7.8 Open Research Challenges

Despite significant advancements, MIMO (Multiple-Input Multiple-Output) antenna technology faces several open research challenges. Achieving compact and efficient designs for MIMO antennas in constrained spaces, such as mobile devices, remains difficult due to mutual coupling and limited physical separation between elements. Enhancing isolation between antenna elements without compromising performance is a persistent challenge. Furthermore, the integration of MIMO antennas with emerging technologies like millimeter-wave (mm Wave) and terahertz (THz) frequencies introduces new complexities, including high path loss and fabrication constraints. Developing reconfigurable and adaptive MIMO systems to support dynamic wireless environments and addressing energy efficiency, cost, and scalability for large-scale MIMO arrays are key areas requiring innovative solutions. These challenges underscore the need for continued research to fully exploit MIMO's potential in next-generation wireless communication systems.

7.9 Summary

This chapter provides details of MIMO antenna for telemedicine applications. MIMO antenna has been designed using FSS and hexagonal shape. The performance of MIMO antenna has been observed. Antenna technology is crucial for enhancing telemedicine applications by providing high-speed, reliable, and low-latency wireless communication. It enables seamless data transmission for remote patient monitoring, real-time video consultations, and emergency medical services. The design of a MIMO antenna for telemedicine must focus on compactness, high efficiency, and minimal interference, ensuring optimal bandwidth, gain, and isolation. Operating in frequency bands like Wi-Fi (2.4 GHz), 5G (Sub-6 GHz, mmWave), or UWB, these antennas improve signal reliability while maintaining energy efficiency for wearable and portable medical devices. By minimizing mutual coupling and optimizing performance through advanced design techniques, MIMO antennas play a critical role in ensuring the success of modern telemedicine systems.

References

1. A. Haleem, M. Javaid, R. P. Singh, and R. Suman, "Telemedicine for Healthcare: Capabilities, Features, Barriers, and Applications," 2021
2. A.V. Lumina, M. Sangeetha, 5G antenna design methods and performance enhancement techniques: a review, in *2023 International Conference on Recent Advances in EElectrical, EElectronics, UUbiquitous CCommunication, and Computational Intelligence, RAEEUCCI 2023*, (Institute of Electrical and Electronics Engineers, 2023). https://doi.org/10.1109/RAEEUCCI57140.2023.10134463
3. Y. Rahayu, T. Kirana, Flexible multi-layer Condura fabric ultra wide-band antenna for telemedicine application, in *2021 9th International Conference on Information and Communication Technology, ICoICT 2021*, (Institute of Electrical and Electronics Engineers, 2021), pp. 133–137. https://doi.org/10.1109/ICoICT52021.2021.9527521
4. M.I. Mapa, D.K.M. Ibarlin, E.R. Arboleda, Optimizing antenna performance: a review of multiple-input multiple output (MIMO) antenna design techniques. Int. J. Sci. Res. Arch. **12**(2), 483–444 (2024). https://doi.org/10.30574/ijsra.2024.12.2.1252

5. M.B. Janjua, A.E. Duranay, H. Arslan, Role of wireless communication in healthcare system to cater disaster situations under 6G vision. Front. Commun. Netw. **1**, 610879 (2020). https://doi.org/10.3389/frcmn.2020.610879
6. D. Li, *5G and Intelligence Medicine—How the Next Generation of Wireless Technology will Reconstruct Healthcare?* (Oxford University Press, 2019, December 01). https://doi.org/10.1093/pcmedi/pbz020
7. P. Sharma, R.N. Tiwari, P. Singh, P. Kumar, B.K. Kanaujia, *MIMO Antennas: Design Approaches, Techniques and Applications* (2022). https://doi.org/10.3390/s22207813
8. W. Wang, Z. Fang, K. Tang, X. Wang, Z. Shu, Y. Zheng, Gain-enhanced wideband antenna sensor integrated with CMOS-based transceiver Chip for human respiratory monitoring in telemedicine diagnosis, in *2022 IEEE MTT-S International Microwave Biomedical Conference, IMBioC 2022*, (Institute of Electrical and Electronics Engineers, 2022), pp. 105–107. https://doi.org/10.1109/IMBioC52515.2022.9790161
9. F.V. Jayasudha, I.R. Sheeba, K. Srilatha, I.M.S. Sanju, P. Chitra, Designing an wearable antenna for telemedicine applications, in *Proceedings of the 2024 10th International Conference on Communication and Signal Processing, ICCSP 2024*, (Institute of Electrical and Electronics Engineers, 2024), pp. 1291–1296. https://doi.org/10.1109/ICCSP60870.2024.10543690
10. A. Singh, A. Kumar, B.K. Kanaujia, FSS inspired two port CP MIMO antenna with enhanced gain for X-band applications. Opt. Commun. **566**, 130698 (2024)
11. J. A. Peral-rosado, R. Raulefs, J. A. López-salcedo, G. Seco-granados, and S. Member, Survey of Cellular Mobile Radio Localization Methods : from 1G to 5G, 20(2), pp. 1124–1148, 2018
12. N. Bhandari, S. Devra, K. Singh, Evolution of cellular network: from 1G to 5G. Int. J. Eng. Tech. **3**(5), 98–105 (2017) http://oaji.net/articles/2017/1992-1515158039.pdf
13. T. Wu, L. Yuan, A. Zhou, Antenna selection technology research in massive MIMO system, in *2022 IEEE Asia-Pacific Conference on Image Processing, Electronics and Computers, IPEC 2022*, (Institute of Electrical and Electronics Engineers, 2022), pp. 967–970. https://doi.org/10.1109/IPEC54454.2022.9777612. II. Antenna design and principle of operation
14. M. Shyam, N. Pavithra, P. Priyadarshini, S. Sakthi Priya, Photo paper based microstrip patch antenna for telemedicine application, in *Proceedings of International Conference on Technological Advancements in Computational Sciences, ICTACS 2022*, (Institute of Electrical and Electronics Engineers, 2022), pp. 396–402. https://doi.org/10.1109/ICTACS56270.2022.9987831
15. A. Atuwayr, A. Asmeida, H.S. Ben Abdelmula, A.A. Anaiba, Optimization of MIMO antenna approach for 5G wireless applications, in *2023 IEEE 11th International Conference on Systems and Control, ICSC 2023*, (Institute of Electrical and Electronics Engineers, 2023), pp. 647–653. https://doi.org/10.1109/ICSC58660.2023.10449883
16. G. Jayaraman, I. Singh, D.K. Choudhary, Gain and isolation improvement techniques for MIMO antenna: Compendious survey. Results Eng. **25**(35), 104482 (2025)
17. J. Ren, W. Hu, Y. Yin, R. Fan, Compact printed MIMO antenna for UWB applications. IEEE Antenna. Wirel. Propag. Lett. **13**, 1517–1520 (2014). https://doi.org/10.1109/LAWP.2014.2343454
18. M.A. Abbas, A. Allam, A. Gaafar, H.M. Elhennawy, M.F.A. Sree, Compact UWB MIMO antenna for 5G millimeter-wave applications. Sensors **23**(5), 2702 (2023). https://doi.org/10.3390/s23052702
19. G. Jayaraman, I. Singh, D.K. Choudhary, Compact monopole MIMO antenna with SRR and DGS for satellite communication, in *2024 IEEE Microwaves, Antennas, and Propagation Conference (MAPCON)*, (IEEE, 2024)
20. L. Malviya, R.K. Panigrahi, M.V. Kartikeyan, MIMO antennas with diversity and mutual coupling reduction techniques: a review. Int. J. Microw. Wirel. Technol. **9**(8), 1763–1780 (2017)

Chapter 8
Ultra-Wideband (UWB) Antennas for Medical Imaging and Sensing

Contents

8.1 Introduction

Ultra-wideband (UWB) technology has emerged as a pivotal component in various fields, notably in medical imaging and sensing applications [1]. The frequency range of 3.1–10.6 GHz allocated for UWB systems by the Federal Communications Commission (FCC) provides immense potential for high-resolution imaging and data communication [2]. The inherent characteristics of UWB signals, such as their ability to offer fine temporal resolution, low-power density, and capacity for high data rates, make them particularly suitable for applications that require precise detection of small changes in dielectric properties, as seen in medical scenarios [3]. UWB antennas are crucial in this domain, as they allow for the transmission and reception of short pulses, which can penetrate biological tissues to identify abnormalities such as tumors [4]. The high bandwidth enables a fine distinction between healthy and cancerous tissues, offering improved diagnostic accuracy. In addition to medical imaging, UWB systems are extensively used in ground-penetrating radars

© The Author(s), under exclusive license to Springer Nature Switzerland AG 2026
A. Segun Ayokunle et al., *Biomedical Antenna Technology*,
https://doi.org/10.1007/978-3-031-95268-5_8

(GPR) for the detection of buried objects and through-wall imaging systems in security and law enforcement [5–7].

The medical imaging application, particularly in breast cancer detection, utilizes UWB technology due to the significant contrast in dielectric properties between normal and cancerous tissues [8]. The wide bandwidth and ability to send and receive ultrashort pulses of UWB antennas allow for detailed internal body imaging, highlighting its importance in non-invasive diagnostics. Moreover, UWB systems are also highly resilient to multipath effects, making them reliable in complex environments where signal reflections might otherwise affect performance [9, 10]. The various application of UWB is shown in Fig. 8.1.

Ultra-wideband (UWB) technology, operating in the range of 3.1 GHz to 10.6 GHz, has revolutionized medical imaging and sensing thanks to the performance of radiating high-resolution, low-power signals. The allocated frequency was, in fact, designed by the Federal Communications Commission (FCC) to offer UWB considerable bandwidth, supporting fine temporal resolution as well as robust data communications. Therefore, UWB is of excellent utility in applications requiring precise detection, including medical diagnostics such as separating types of tissues. It has found that high data rate capacity, short pulse transmission, and multipath interference are characteristics which find UWB to be superior in sensitive applications like medical imaging, security, and ground-penetrating radar [11]. The benefits of UWB in medical imaging find their source through ultrashort pulses at a wide range of frequency. This guarantees high spatial and temporal resolution needed to identify fine structures and slight changes within biological tissues. In this regard, UWB antennas emitting such pulses can attain the depth needed for penetration through tissues while maintaining the detail level to distinguish between healthy and abnormal regions. To be more specific, in the detection of breast cancer, UWB can exploit variations in dielectric properties between cancerous and normal tissues

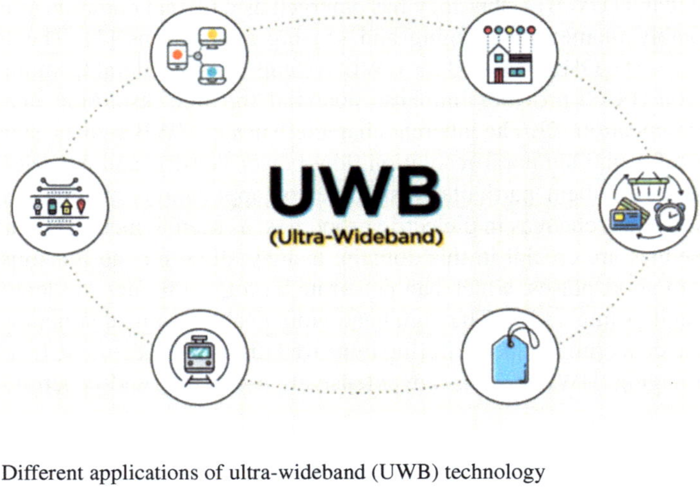

Fig. 8.1 Different applications of ultra-wideband (UWB) technology

to make early-stage identification easier by non-invasive high-resolution imaging [12].

Because the power density of UWB technology is low, it minimizes interference with other hospital devices and thereby ensures operation within the delicate medical environment. Moreover, the high data rate of UWB makes imaging and processing rapid and can lead to real-time diagnostic capabilities, which is highly valued in fast-paced environments. Antennas for UWB are a key aspect of the technology and hence play a crucial role in the efficiency of UWB technology in medical applications, transmitting and receiving the high-resolution pulses on which UWB imaging relies. Short pulse duration is critical in preventing signal blurring, a major drawback experienced in most other imaging techniques, and ensures clearer images. In such applications as breast cancer imaging, any early detection of small tumors leads to greatly improved patient outcomes (Fig. 8.2).

The UWB is also helpful in the non-invasive diagnostic techniques. While the conventional imaging methods like the X-rays are based on the ionizing radiations, exposure for long periods to this kind of radiation may prove hazardous and must be made sparingly. In contrast, UWB makes use of non-ionizing radiation and is, therefore, safer for periodic diagnoses. Moreover, because of its nature, multipath effects, which normally lead to an interference signal, do not degrade the images much [14]. UWB apart from medical purposes has lately been caught on the security and GPR areas. For instance, in through-wall imaging, it is because of this UWB pulses penetrability that teams of security can identify movements and objects that are secreted behind a wall [15]. In GPRs, UWB pulses penetrate the ground and thus facilitate the detection of items such as pipelines, landmines, and archaeological artifacts. Conversely, the high resolution is rather helpful for construction, geological studies, and archaeology as subsurface detection is key.

Ever since, the advancement in the design of antennas, signal processing, and material engineering contributed significantly to the enhancement of UWB technology and augmented its applications, ranging from medical to other non-medical applications. The latest research is now rapidly concentrating on making UWB antennas much more compact and flexible for implantation within wearable health

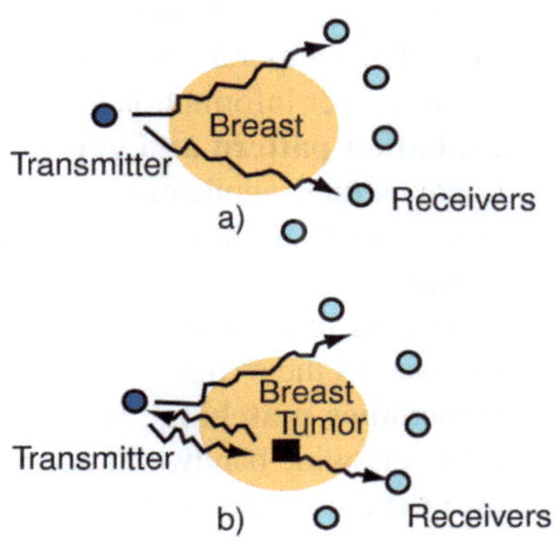

Fig. 8.2 Basic microwave imaging problem. (**a**) Reflected waves from breast without a tumor. (**b**) Reflected waves from breast with a tumor [13]

monitors, which would allow constant, real-time monitoring of health status. These developments also make possible noncontact vital sign monitoring—that is, the measurement of heart rate and respiration, among others—thereby providing very useful applications in intensive care units or for monitoring patients from a distance. In security and industrial environments, UWB is being utilized to provide location with accuracy to the centimeter level for highly accurate indoor positioning systems used to track individuals or equipment in environments where people cannot easily be seen, such as warehouses or underground. The hope of UWB technology is the never-ending development of a sphere of applications, gradually moving on to increasingly complex tools with answers to challenges in health, safety, and other spheres [15, 16]. The powerful potential of UWB technology on medical imaging and security applications lies in its fine resolution, relatively low interference, and non-invasive nature. Advancement in UWB technology will help further innovate healthcare, public safety, and industries; it will further advance more accurate diagnostics and reliable detection of diverse fields.

8.2 UWB Antenna Principles and Characteristics

The fundamental principles of UWB antennas are rooted in their ability to transmit and receive very short pulses of electromagnetic waves over an exceptionally broad frequency range. Unlike narrowband antennas that operate within a confined spectrum, UWB antennas cover a wideband range from 3.1 GHz to 10.6 GHz. This broad frequency coverage ensures that UWB antennas can support applications that require high data transfer rates and fine temporal resolution. Ultra-wideband antennas are designed to operate in a very wide frequency spectrum where its input bandwidth is greater than 1 GHz, or as much as 10 GHz [17]. It is capable of providing a large data rate with high precision. UWB antennas are especially critical to use in applications such as medical imaging, remote sensing, and radar, where the applications require precision and accuracy.

1. **Wideband frequency operation**: The UWB antennas operate over a wide frequency range that makes it possible to create short pulses to generate high spatial as well as temporal resolution. It is very useful for applications related to imaging wherever information about the structure at high resolution is needed.
2. **Radiation pattern and gain:** UWB antennas can be omnidirectional or directional as their applications suggest. For example, an omnidirectional pattern is pretty much applicable to catching signals from many directions, whereas a directional pattern radiates more energy concentrated in a smaller region, thus increasing resolution in that region. To ensure consistent performance at stable gain across the frequency range.
3. **Impedance matching:** Since impedance matching is excellent throughout the relatively wide bandwidth of operation—often return loss is better than −10 dB over the entire bandwidth—there is essentially no reflected power and therefore

high efficiency. Tapered feeds, stub matches, or broadband impedance matching circuits can be used.
4. **Pulse fidelity:** To have high performance by UWB antennas for time domain applications, the pulses that are transmitted and received should be with minimal distortion. That ensures clear and sharp images in medical and sensing applications.
5. **Material and size consideration:** Low-loss high-dielectric material is mostly used for UWB antennas: for instance, Rogers, Teflon, etc., to make lesser losses and improve the antenna designs. In compact designs, fractal geometries or substrate integration techniques are used to miniaturize the antenna.

8.3 UWB Antenna Design for Imaging and Sensing Devices

Ultra-wideband (UWB) medical imaging and sensing antennas will be optimized to be compact to ensure achieving a high spatial resolution with high sensitivity, thereby allowing detailed collection of biological information. These antennas operate in the frequency range 3.6 GHz to 10 GHz, ideally balancing the desired depth of tissue penetration with the actual spatial resolution that is necessary to obtain good quality medical imaging. The selected range of frequency supports fine resolutions in imaging biological tissue and such applications require medical tumor detection, organ health monitoring, and physiological change measurement [18].

Common UWB geometries for medical imaging range from planar monopoles through bow-tie and Vivaldi type to build big bandwidths as well as diversity in radiation characteristics. Fractal or meandering techniques keep the performance of the antennas at optimum levels in compact devices used in medicine. Stability gain across the frequency band is a requirement; omnidirectional patterns of radiation support multiple-angled signal capture, while directional patterns help focus on particular regions of radiation with better resolution. Proper impedance matching in the design is also important to ensure that no power loss and signal degradation occur [19].

UWB antennas radiate short pulses with high susceptibility to distortion. Hence, pulse fidelity is of paramount importance to ensure that received signals are distortion less and error free. SAR compliance is also crucial as it limits the absorption of excessive energy by biological tissues, thus keeping the patient inside the scanner safe from harm. When high-quality, low-loss substrates like Rogers and Teflon are used, maximum efficiencies are achieved, and overall antenna sizes are minimized.

It contains simulation tools like CST software, where bandwidth, gain, and SAR can be optimized with accurate performances. It enables the development of advanced UWB antennas with their capabilities to perform high-resolution medical imaging and real-time monitoring; hence, they are expanding innovations in healthcare technology. Single element UWB antenna design is shown in Fig. 8.3. Different view of a two-element UWB MIMO antenna design is shown in Fig. 8.4.

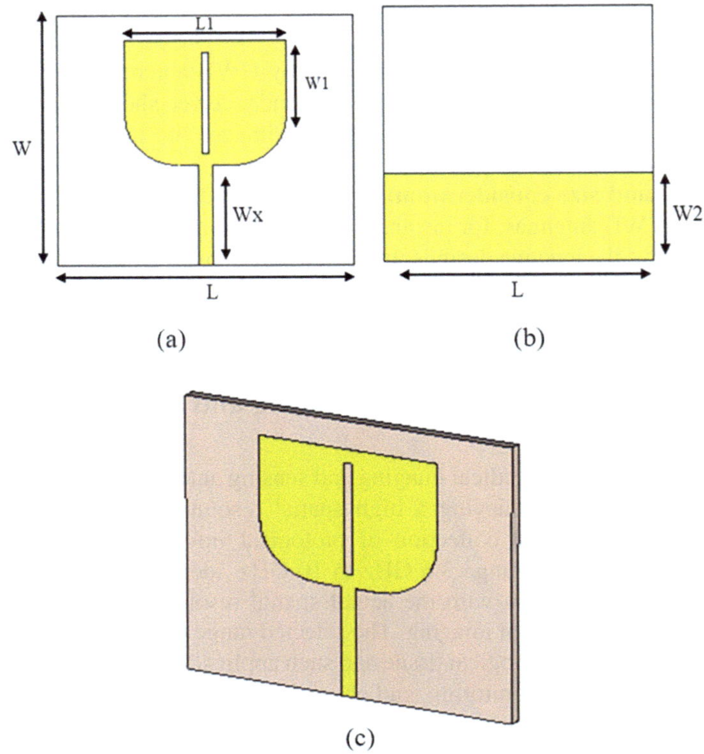

Fig. 8.3 Configuration of single element UWB antenna (**a**) front view, (**b**) back view, and (**c**) prospective view

8.4 Problem Formulation

The aim of this chapter is to design a compact UWB MIMO antenna for medical imaging and sensing applications. Target frequency range is 3.6 GHz to 10.6 GHz. High spatial and detail resolution within biological tissues.

Design Requirements

1. *Broad frequency bandwidth:* Ultra-wideband operation in the given band should allow short pulses, which helps in achieving high spatial and temporal resolutions, to be sent and received.
2. *Compactness, lightweight*: The antenna needs to be compact and lightweight in a sense that it does not compromise the functionalities of the wearable or portable devices.
3. *Radiation pattern and gain stability*: The antenna should have stability in gain over the bandwidth with omnidirectional or directional radiation patterns as suits a certain imaging or sensing application.

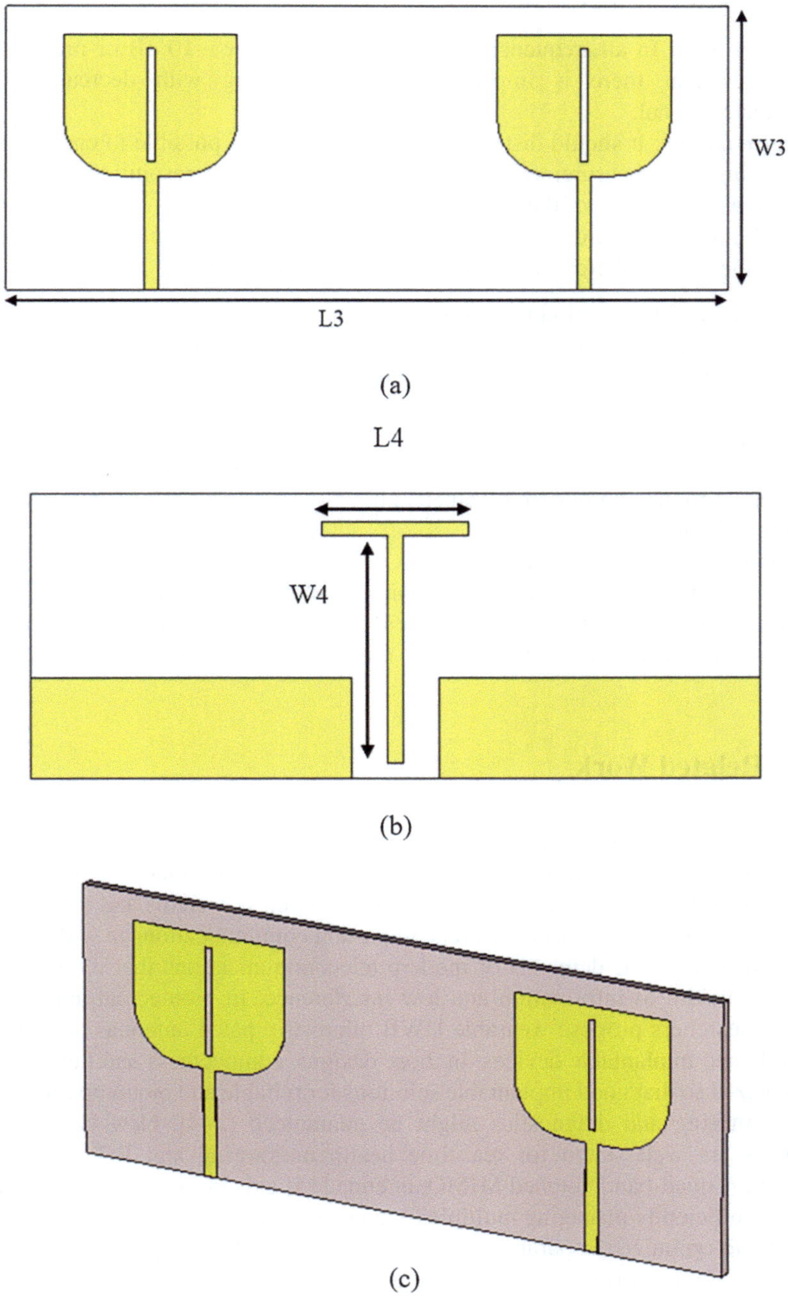

Fig. 8.4 Configuration of two element UWB MIMO antenna (**a**) front view, (**b**) back view, and (**c**) prospective view

Performance Metrics

1. *Return loss:* In all frequencies, return loss should be −10 dB or more, which secures that there is impedance matching along with decrease of the reflected signal.
2. *Pulse fidelity*: It should distort pulses as minimum as possible to ensure proper imaging and measurement. Distorted pulses decrease resolution and further decrease the clarity of the image.
3. *SAR:* Safety standards compliance. It will result in minimum energy absorption into the tissues, thereby minimizing adverse thermal effects.

Material and Fabrication Constraints.

Low-loss and high-dielectric materials, such as from Rogers or Teflon, should be used to minimize power loss and achieve optimal performance in the very compact structure of the antenna.

Simulation and Testing.

Testing or equivalents of optimization and CST electromagnetic simulation tool with regard to the appropriate parameters like S-parameters, gain, efficiency, SAR, and the radiation pattern.

Application-Specific Requirements.

The antenna will be tailored toward high-resolution tissue imaging or real-time physiological monitoring, such as heartbeats and respiration sensing, with emphasis on sensitivity, accuracy, and reliability in clinical/diagnostic environments.

8.5 Related Work

Recent developments on ultra-wideband antennas focus attention on compact, flexible, and multiband solutions for widely diverse applications. One design introduces a super-UWB monopole antenna for multiband wireless systems and demonstrates UWB's potential to support interference-resistant communications in wide frequencies, which suits the demands of modern telecommunications that require both a high bandwidth of information and low interference. In biomedical applications, some researchers propose wearable UWB microstrip patch antennas designed for on-body and implantable devices. In those designs, compactness and flexibility are emphasized so that good implantable solutions for reliable and biocompatible medical monitoring and diagnostics might be guaranteed [2–4]. New band-notched antennas are well suited for real-time health monitoring and IoT applications [15–19]. A quad-band notched MIMO antenna [15] aids in improving IoT connectivity by efficiently managing multiple frequencies; others [19] instead use machine learning to optimize the parameters involved to ensure stable operation across 3G, LTE, and 5G networks, thus increasing adaptability and efficiency in smart wearables. UWB phased array antennas also offer high-resolution imaging and accurate detection; hence, they are pretty suitable for applications in the areas of radar and surveillance [26]. These arrays utilize the wide bandwidth offered by UWB to achieve superior spatial resolution and precision object detection. To address the issue of cost-effectiveness, some authors present low-profile planar patch antennas that are designed with scalability in mind for use in consumer electronics, where size and budget play a considerable role. Also, photonic crystal antennas enable multiband 5G applications with improved data rates and efficient multiband performance. Collectively, these developments bring into focus the ability of UWB antennas to be adopted suitably within wearable tech, IoT, telecommunication, and radar applications [24].

8.6 Simulation Results and Discussion

See Figs. 8.5, 8.6, 8.7, 8.8, 8.9.

8.7 Open Research Challenges

The challenges will vary with the application, especially for wearables, medical, and IoT applications. Miniaturization plays a very important role in such designs. Miniaturized configurations are desired but need to offer important performance in terms of gain, efficiency, and bandwidth. This is particularly pertinent in portables and wearable devices due to the space constraint. Another key problem area is in modelling signal propagation—understanding how UWB signals behave in complex, dynamic environments, for example, medical or indoor deployments. Much

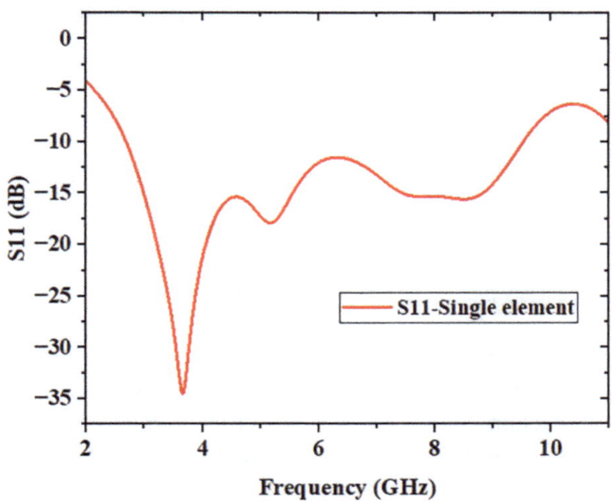

Fig. 8.5 Return loss vs frequency plot for single element antenna

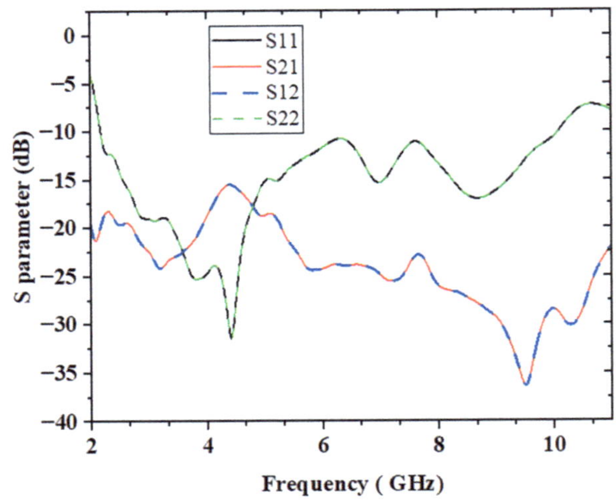

Fig. 8.6 Scattering parameters vs frequency plot for UWB two-port MIMO antenna

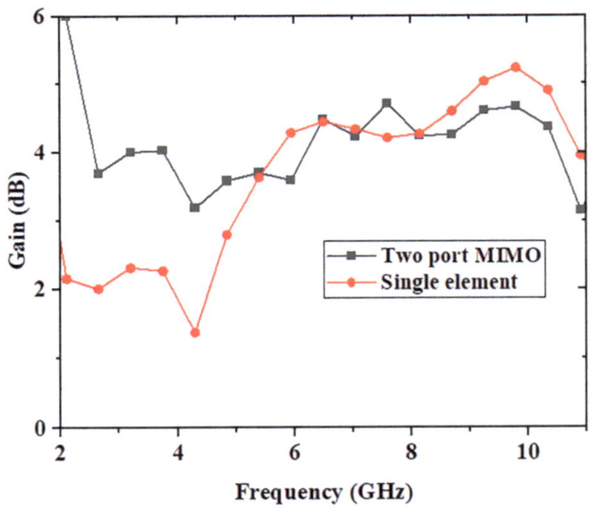

Fig. 8.7 Gain vs frequency plot for UWB two-port MIMO antenna

higher signal reliability, fidelity, and precision stem from better modelling. Another important area is interference mitigation since UWB systems are meant to coexist with other wireless technologies. The designs or techniques of interference resilient designs are pertinent for smooth coexistence and sustainment of the performance. Research is in progress on material and structural optimization to find materials that are lightweight yet flexible and efficient for UWB antennas, so they can be

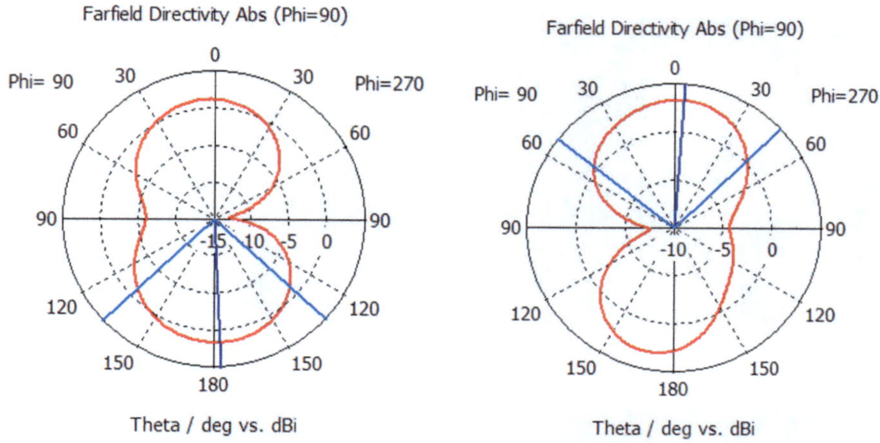

Fig. 8.8 Radiation pattern at frequency 2.4 GHz and 5 GHz for UWB two-port MIMO antenna

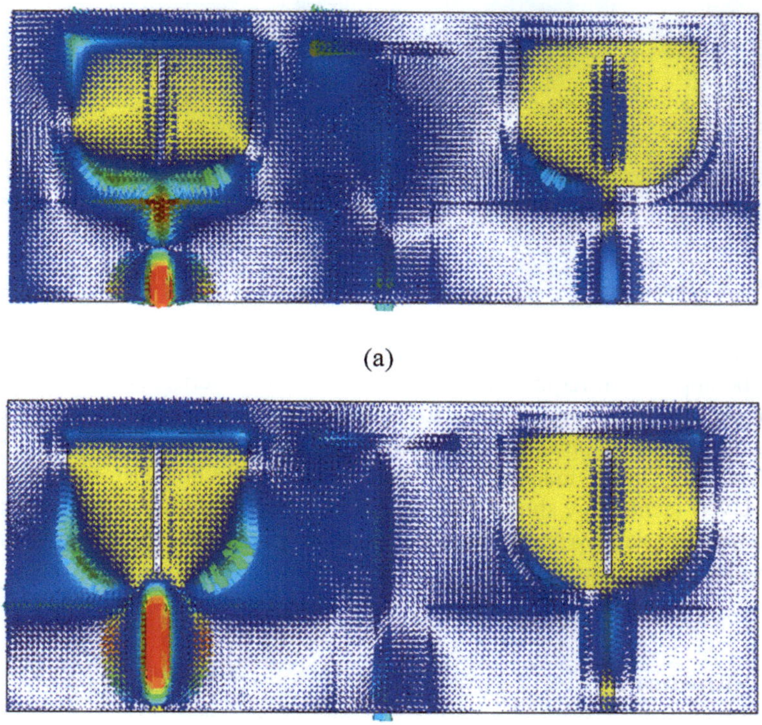

(a)

(b)

Fig. 8.9 The current density distribution is recorded for various phases such as 0°, 90°, 180°, and 270° at frequency 5GHz for UWB two-port MIMO antenna

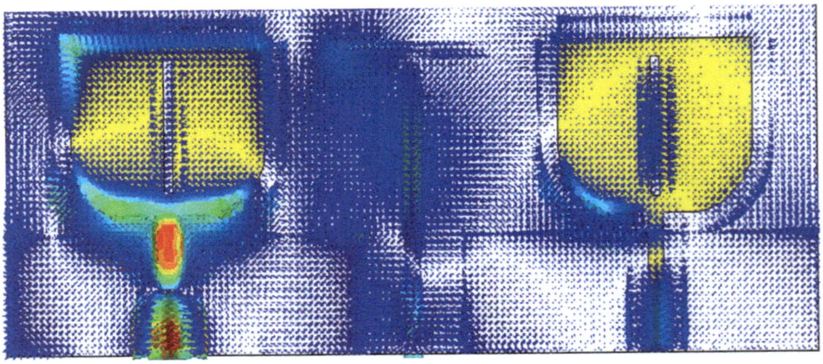

(c)

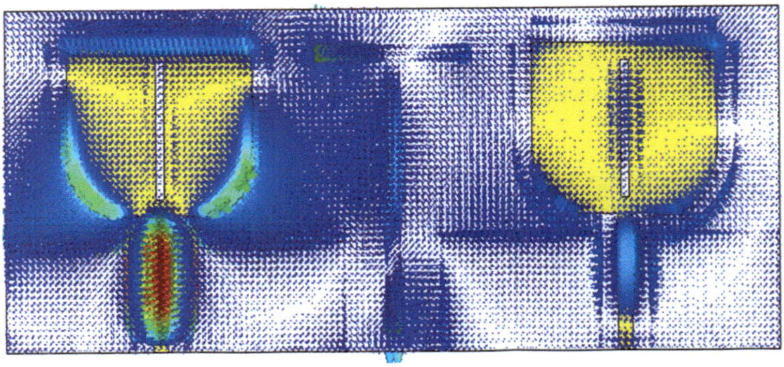

(d)

Fig. 8.9 (continued)

mass-produced with comfort. Energy efficiency is the biggest hurdle, especially when the application requires extended lifetimes on batteries, such as continuous health monitoring and other remote IoT systems. This is where the challenge comes in—to further reduce power consumption without detractions to signal quality—an area that is very much under active development. Last but not least, standardization of the UWB technology should have to ensure interoperability, compliance and in one, uniformity across a myriad of applications in order for the technology to mature in these various industries.

8.8 Summary

The ultra-wideband antenna presented in this chapter is simulated through CST Studio Suite for various applications such as medical, sensing, and imaging purposes. The steps of optimization with high efficiency and stable gain through impedance matching over the frequency range of 3.1 to 10.6 GHz with return loss kept below −10 dB involve three steps which include boundary conditions, feeding port assignment, and S-parameter simulation for the optimization process. Further study of the proposed design involved gain, efficiency, and radiation patterns for maximum penetration in tissue with clear imaging. Further optimization of patch size and feed position parameters boosts performance further. It has some issues on miniaturization, reduced interference, and biocompatibility that require further developmental work.

References

1. U. Rafique, S. Pisa, R. Cicchetti, O. Testa, M. Cavagnaro, Ultra-wideband antennas for biomedical imaging applications: a survey. Sensors (Basel) **22**(9), 3230 (2022). https://doi.org/10.3390/s22093230
2. E.G. Lim, Z. Wang, C.U. Lei, Y. Wang, K.L. Man, Ultra-wideband antennas: past and present. IAENG Int. J. Comput. Sci. **37**(3) (2010)
3. M.M. Salah, M. Hossam, H.A. Ragheb, A.R. Eldamak, Efficient super-ultra-wideband monopole antenna design for multi-band wireless applications, in *2024 International Telecommunications Conference (ITC-Egypt)*, (2024), pp. 531–536
4. B. Khokher, A.B. Gurulakshmi, C.J. Arpitha, D.G. Bhat, A.B. Patil, B.L. Patra, A comprehensive study of wearable microstrip patch antennas for biomedical applications, in *2024 International Conference on Computational Intelligence for Green and Sustainable Technologies (ICCIGST)*, (2024), pp. 1–6
5. R. Parveen, A. Debnath, F.S. Abir, M. Rahman, A slotted monopole patch antenna design for ultra-wideband wireless applications, in *2023 IEEE International Conference on Telecommunications and Photonics (ICTP)*, (2023), pp. 1–5
6. R. Vanitha, S. Mary Praveena, A flexible wearable antenna on-body and implant communications using UWB technology, in *2023 International Conference on New Frontiers in Communication, Automation, Management and Security (ICCAMS)*, vol. 1, (2023), pp. 1–4
7. Sneha, P.K. Malik, A. Gehlot, A key-shaped ultra-wideband antenna for IoT applications, in *2023 International Conference on Artificial Intelligence and Smart Communication (AISC)*, (2023), pp. 504–508
8. S. Sharma, M. Kumar, H. Nigam, M. Mathur, Dual band circular patch coplanar microstrip patch antenna for X band and Ku band applications, in *2021 IEEE Indian Conference on Antennas and Propagation (InCAP)*, (2021), pp. 344–346
9. M. Rahman, F. Ahmed, A. Md, A. Rahman, Low-cost low-profile planar patch antenna for ultra-wideband applications, in *2019 2nd International Conference on Innovation in Engineering and Technology (ICIET)*, (2019), pp. 1–4
10. H.N. Awl, B.A. Karim, Y.I. Abdulkarim, L. Deng, D.A. Hassan, M. Karaaslan, Broadband microstrip antenna for C-band, X-band, and KU-band applications, in *2019 International Conference on Information and Telecommunication Technologies and Radio Electronics (UkrMiCo)*, (2019), pp. 1–5
11. X. Zhao, S. Huang, Z. Jiang, Analysis and Design of a Novel Ultra Wideband Planar Antenna, in *2019 IEEE International Conference on Computational Electromagnetics (ICCEM)*, (2019), pp. 1–3
12. S. Chakraborty, A.J. Islam, M.M. Farhad, M.M. Hasan, M.S.B. Nesar, M.A. Rahaman, Performance studies of UWB microstrip antenna for multipurpose biotelemetry applications, in *2018 International Conference on Innovations in Science, Engineering and Technology (ICISET)*, (2018), pp. 322–327

13. E.C. Fear, P.M. Meaney, M.A. Stuchly, Microwaves for breast cancer detection? IEEE potentials **22**(1), 12–18 (2003)
14. S. Doddipalli, A. Kulkarni, P. Kaole, A. Kothari, Slotted substrate miniaturized ultra wideband antenna for WBAN applications, in *2018 9th International Conference on Computing, Communication and Networking Technologies (ICCCNT)*, (2018), pp. 1–4
15. B. Jeyapoornima, J.J.J. Sheela, UWB band-based patch antenna for wireless communication systems, in *2023 7th International Conference on Electronics, Communication and Aerospace Technology (ICECA)*, (2023), pp. 1662–1665
16. H.M.Q. Rasheda, N. Abdullah, Q. Abdullah, N. Farah, A. Uğurenver, A. Salh, A.R.O. Mumin, Design of UWB antenna for microwave imaging using modified fractal structure, in *2021 International Congress of Advanced Technology and Engineering (ICOTEN)*, (2021), pp. 1–4
17. W. Xu, J. Lin, J. Wang, H. Zhang, A novel ultra-wideband microstrip-to-waveguide transition based on microstrip tramped antenna, in *2018 International Conference on Microwave and Millimeter Wave Technology (ICMMT)*, (2018), pp. 1–3
18. J. Chu, D. Chengzhu, R. Yang, A wearable CPW-fed quad-band-notched two-port MIMO antenna for IoT applications, in *2024 International Applied Computational Electromagnetics Society Symposium (ACES-China)*, (2024), pp. 1–3
19. U.R. Iman, M. Zada, A. Basir, S. Hayat, Y.-H. Lim, H. Yoo, IoT-enabled real-time health monitoring via smart textile integration with LoRa technology across diverse environments. IEEE Trans. Industr. Inform. **20**(11), 12803–12813 (2024)
20. X.Y. Liu, R.X. Zhang, H.R. Zu, R.G. Song, D.P. He, A conformal tri-band band notched antenna for WLAN/WiMAX applications, in *2024 IEEE International Workshop on Radio Frequency and Antenna Technologies (iWRF&AT)*, (2024), pp. 315–317
21. S.A. Babale, T.K. Geok, S.K.A. Rahim, C.P. Liew, U. Musa, M.F. Hamza, Y.A. Bakhuraisa, L.L. Lim, Machine learning-based optimized 3G/LTE/5G planar wideband antenna with tri-bands filtering notches. IEEE Access **12**, 80669–80686 (2024)
22. L. Zhao, Y. Wang, C. Liu, D. Song, H. Chufeng, C. Li, H. Zhao, Z. Wang, Compact circular-shaped MIMO antenna covers UWB bandwidth with four frequently-used band-notched characteristics for multi-scenario applications. IEEE Access **12**, 32762–32771 (2024)
23. A.D. Prasetyo, T. Yunita, D. Hamdani, A. Munir, Control points rearrangement of spline-based CPW-fed spearhead monopole antenna using particle swarm optimizer, in *2024 Photonics & Electromagnetics Research Symposium (PIERS)*, (2024), pp. 1–5
24. M.N. Moussa, M.A. Madi, K.Y. Kabalan, Breast tumor detection, sizing and localization using a 24-element antenna Array. IEEE J. Biomed. Health Inform. **26**(10), 5109–5121 (2022)
25. M.N. Moussa, M.A. Madi, K.Y. Kabalan, Miniaturized inward fractal antenna for breast cancer detection, in *2021 International Conference on Electrical, Computer and Energy Technologies (ICECET)*, (2021), pp. 1–4
26. H. Kaschel, C. Ahumada, Design of a microstrip antenna for breast cancer detection, in *2021 IEEE CHILEAN Conference on Electrical, Electronics Engineering, Information and Communication Technologies (CHILECON)*, (2021), pp. 1–5
27. N. Bagheri, E. Teixeira, F.J. Velez, J.M. Peha, Multi-band resonant photonic crystal antenna for 5G applications, in *2024 IEEE 22nd Mediterranean Electrotechnical Conference (MELECON)*, (2024), pp. 526–531
28. V.G. Sawant, P. Kadam, V. Mangrulkar, H. Gawade, Y. Patil, 5.8GHz ISM band low cost button antenna for smart wearable application, in *2023 6th International Conference on Advances in Science and Technology (ICAST)*, (2023), pp. 638–642
29. R. GHolami, H. Mehrpourbernety, B. Zakeri, UWB phased array antennas for high resolution radars, in *2013 International Symposium on Electromagnetic Theory*, (2013), pp. 532–535
30. H.M. Bernety, R. Gholami, B. Zakeri, M. Rostamian, Linear antenna array design for UWB radar, in *2013 IEEE Radar Conference (RadarCon13)*, (2013), pp. 1–4
31. M. Mokhtaari, J. Bornemann, Printed-circuit antennas for 3–30 GHz and 3–60 GHz UWB applications, in *2010 Asia-Pacific Microwave Conference*, (2010), pp. 1922–1925

Chapter 9
Reconfigurable Antennas for Adaptive Healthcare Devices

Contents

9.1 Introduction

In recent years, the reconfigurable antenna has emerged to play a pivotal role in adaptive healthcare devices. Conventional antennas, however, struggle to meet the miniaturization demands. The reconfigurable antennas promise efficient designs by varying their frequency, radiation pattern, or polarization in response to changing environments. This makes the reconfigurable antenna ideal for applications where multiple antennas are needed to be installed in a limited space. The reconfigurable antennas satisfy the space constraints in the transmitter and receiver devices. For the medical and healthcare applications, the size of the device is a significant constraint. Implant devices and wearable technology require miniaturized medical devices.

© The Author(s), under exclusive license to Springer Nature Switzerland AG 2026
A. Segun Ayokunle et al., *Biomedical Antenna Technology*,
https://doi.org/10.1007/978-3-031-95268-5_9

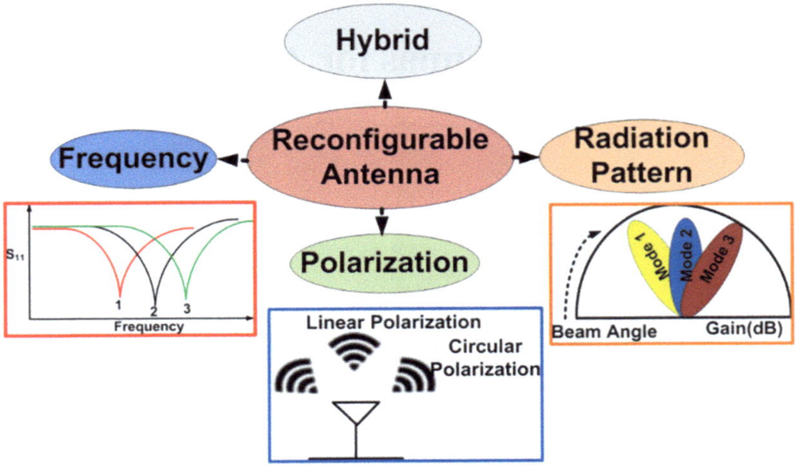

Fig. 9.1 Reconfigurable antenna [9]

Therefore, the reconfigurable antennas can be helpful in healthcare applications. A single reconfigurable antenna can fulfil multiple user demands [1] (Fig. 9.1).

The antenna parameters, mainly frequency, pattern, polarization, or a combination of these parameters, are modified dynamically to achieve reconfigurability in the antenna. According to the above-mentioned methodologies, the reconfigurable antennas are broadly categorized into frequency, pattern, polarization, and hybrid. An antenna is tuned to different frequencies using a switch, referred to as a frequency reconfigurable antenna. If the switching mechanism is used to steer the radiation beam of the antenna, it is called a pattern reconfigurable antenna. The antenna has the capability to switch between different polarization types and is said to have polarization reconfigurability. Sometimes, hybrid reconfigurable antennas can also be used. The hybrid scheme is achieved by combining two or all three types of reconfigurability. A single antenna can have both frequency and pattern reconfigurability, frequency and polarization reconfigurability, pattern and polarization reconfigurability, or even frequency pattern and polarization reconfigurability. The reconfigurable antennas have applications across all implantable devices, non-implantable monitoring systems, wearable devices, and medical telemetry devices. The application of reconfigurable antennas is variable and depends upon the surrounding environment and the nature of the application. For example, in the case of implant devices, frequency and pattern reconfigurable antennas are in demand, and in the case of telemetry communication, all frequency pattern and polarization reconfigurable antennas are useful. The wireless body area network is another prominent application of the frequency reconfigurable antenna. [2, 3]

Many switches are available to achieve reconfigurability in antennas, such as PIN and varactor diodes, FET, RF-MEMS, and photodiodes. PIN diode and the

Table 9.1 Antenna design substrate and their feature

Substrate	Feature	Dielectric constant	Loss tangent
Epoxy	Dielectric substrate made from epoxy resin	4.2	0.02
RT Duroid	A high-quality dielectric material with a low dielectric constant and low loss tangent	2.2	0.0009
Polydimethylsiloxane (PDMS)	Flexible, transparent, and elastic substrate made	2.68	0.04
Polyethylene	Flexible, plastic material	2.25	0.001
Polyimide	Flexible, thin sheet made of polyimide material	4.3	0.004
Teflon (polytetrafluoroethylene-PTFE)	Low dielectric constant, flexible, and able to operate in high-temperature environments	2.1	0.001

RF-MEMS are widely used switches in the literature. In most cases, the reconfigurable antennas are implemented on the nonflexible substrates. However, the reconfigurable antennas for wearable applications are also on flexible substrates. Flexible substrates are needed because of the difficulties in making stable connections between rigid electronic components and flexible textile conductors. The popularly used nonflexible and flexible substrates are given in Table 9.1 [4].

The reconfigurable antenna offers multiple advantages over the conventional antenna, such as space requirements and implementation of diversity techniques. Since the different diversity techniques can be implemented with reconfigurable antenna, all the advantages of diversity are available with reconfigurable antennas. The reconfigurable antennas have some limitations and disadvantages, such as increased complexity and degraded performance, compared to the conventional antennas. The advantages and limitations of reconfigurable antennas compared to conventional antennas are given in Table 9.2.

9.2 Basics of Reconfigurable Antennas

The reconfigurable antennas can dynamically vary their parameters according to the need or surrounding environment. This dynamism in the parameter variation improves the performance of reconfigurable antennas and finds broad applications in modern wireless communication systems. The changes in antenna parameters are achieved by adjusting the physical structure, changing materials, or implementing electro-mechanical components [5].

Table 9.2 Comparison between the conventional and reconfigurable antenna

Feature	Conventional antenna	Reconfigurable antenna
Applications	The antenna can accommodate only one frequency/direction of radiation pattern/polarization type in wireless services	The antenna can accommodate multiple services together
Space requirement	Multiple antennas are required for individual frequency bands/the direction of the radiation/the polarization type requires more space Not suitable for MICS and wearable antenna applications	Space requirements are lower than those of conventional antennas
Front-end complexity	Front end is simple. Sometimes, filters can also be designed in the same antenna structure and referred to as filtenna	Generally, the front end is complex. It further increases if the antenna requires the tuning of mechanical switches
RF performance	Offers excellent performance	It is tough to achieve excellent performance in every switching condition. The switches themselves cause noise
Diversity	The individual antenna has to operate at a specific frequency/ direction of radiation/ polarization to achieve diversity	It can provide the frequency, pattern, and polarization diversity easily
Radio coexistence	Strong coupling is available among the radios as there is very little space among antennas	Radio coexistence is rarely possible because only one antenna is operating and that can support only one service at a time
Cost	The cost feature is extremely dynamic. If a single antenna is used, the cost is lower. The use of multiple antennas for an application will increase the cost. Also, the increased number of cable connections will cause higher costs	Here also, the cost is dynamic. One antenna is replacing multiple antennas, so the cost is low. The prices of low-power consumption and low loss switches are high. Many low-cost switches are available, but researchers have to compromise performance

9.2.1 Types of Reconfigurable Antennas

As mentioned above, the reconfigurable antenna can be classified based on three main antenna parameters: operating frequency, radiation pattern, and polarization. The antennas are made adaptive to tuning the different frequencies, steering the angle of the radiation pattern, or switching the type of polarization. One more possibility is that the two or more approaches can be combined and named hybrid reconfigurability. The classification of reconfigurable antennas used in healthcare applications is shown in Fig. 9.2.

(a) **Frequency Reconfigurable Antennas:** The antennas have multiple frequency bands, and the user can change the operating bands by implementing tuning

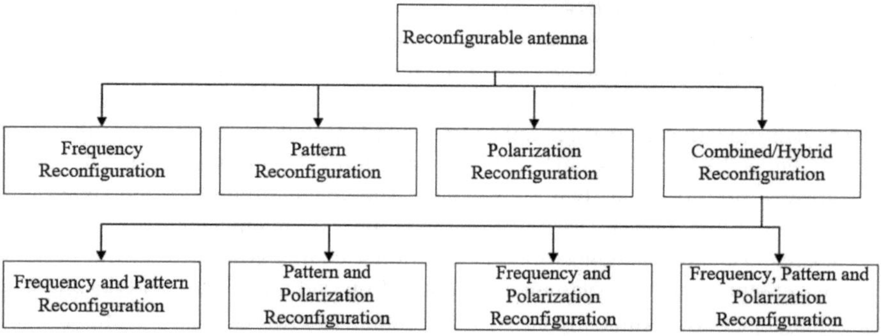

Fig. 9.2 Categorization of reconfigurable antennas

devices like MEMS switch, varactor diode, PIN diode, etc. in the antenna. Since the frequency is the simplest characteristic to change in an antenna, frequency reconfigurability is the most widely used type of reconfigurability. The PIN diode is the commonly used switching device because it offers minimum insertion loss, less cost, excellent isolation, and low-power profile. The frequency reconfigurable antenna is used in multiband applications of RF communication systems. The second popular element for switching in frequency reconfigurable devices is varactor diodes. They offer multiple resonant frequencies by changing the applied voltage and, correspondingly, the capacitance in the antenna element. The frequency reconfigurability is achieved by changing in the size of the antenna physically or electrically. The RF-MEMS is another element used to regulate the antenna size. A low-voltage RF-MEMS is required for frequency tuning. In such cases, the frequency tuning capacity can also be derived using an antenna coil. The size of the reconfigurable system is a challenge here. Micro-fabrication is employed to optimize the size while maintaining performance. Further, it is also possible to use FET as a switching element. The FET offers low insertion loss and does not affect the radiation pattern. Biasing the FET is also easy in such type of switch fabrication on the antenna. The reconfigurability is not limited to the use of a single switch. It is possible to use multiple switching elements that offer different combinations of current flow in the antenna patch, and therefore, multiband operations can be achieved easily [6].

(b) **Pattern Reconfigurable Antennas:** Radiation beam steering is another essential technology that is required in many applications. It allows the antenna to change the direction of the radiation pattern while keeping other properties unchanged. The pattern reconfigurability allows the on-demand change in the radiation direction. High-throughput in data transmission and better coverage are the key advantages achieved with pattern reconfigurable antennas. The major lobe of the radiation is steered in the useful direction, and the interference coming from the other directions is rejected. Therefore, the maximization of energy in/from a particular direction and energy nulling in/from is achieved simultaneously. The SINR performance of the antenna system has improved. A further improvement in performance can be achieved by implementing the phased array system. In the case of the phased array, the steering mechanism is essential. Sometimes, mechanical steering is used in a phased array system. Synchronization in the excitation of individual antenna elements is the other challenge for array designing. The use of multiple beamforming antennas in the array system provides better efficiency. The signal strength of the signal in a particular direction is dependent on the number of antenna elements. However, the use of more antennas causes additional challenges, such as the optimal placement of antenna elements, binding of the radiation, size, and design. The challenges are further increased if the antenna array is used for implant and wearable applications [7].

(c) **Polarization Reconfigurable Antennas** allow users to switch between linear, circular, and elliptical polarization. The antenna feed position and geometrical changes are made to achieve polarization reconfigurability. A polarization reconfigurable antenna with circular or linear polarization offers advantages such as enhanced channel capacity, suppression of multipath, and the possibility of polarization coding. In many research, the design of a microstrip radiator is varied to achieve a polarization reconfigurable antenna. Although the polarization reconfigurability is achieved at a cost of reduced bandwidth. The most promising aspect of the polarization reconfigurable antennas is that they are best used at the receiver, so the improper orientation of the transmitter antenna is neutralized. The feature is particularly advantageous in IMD and wearable applications. Because in such applications, the orientation of the transmitting antenna is dependent on the patient's body movement [8].

One potential application of such an antenna with polarization diversity is the body-centric wireless communication system (BWCS). The concept of BWCS is introduced in paper [3]. The paper suggests two types of communication links: short-range communication links, also called inductive links, and long-range links, also called far-field radio frequency links. In the long-range link, the chances of multipath losses are high as the signal travels a longer distance. Figure 9.3, mentioned below, indicates the basic architecture of BWCS.

(d) **Hybrid Reconfigurability:** The reconfigurability can also implemented by combining multiple features, such as frequency and pattern reconfigurability. The combination of two or more reconfigurability is possible. Out of all frequency and radiation patterns, combining is the most frequent reconfigurability. There are many applications where the switching of multiple frequencies is required along with a specific direction of radiation pattern.

Take the example of frequency reconfigurable antenna used in cognitive radio applications. The antenna needs to switch to different frequencies. If the same antenna is installed at the transmitter of the secondary user (SU) or cognitive radio (CR) node, since it has an omnidirectional radiation pattern, it will radiate equal power in all directions. As the receiver SU is placed in a particular direction, implementing a frequency reconfigurable antenna will waste a lot of radiating power. To

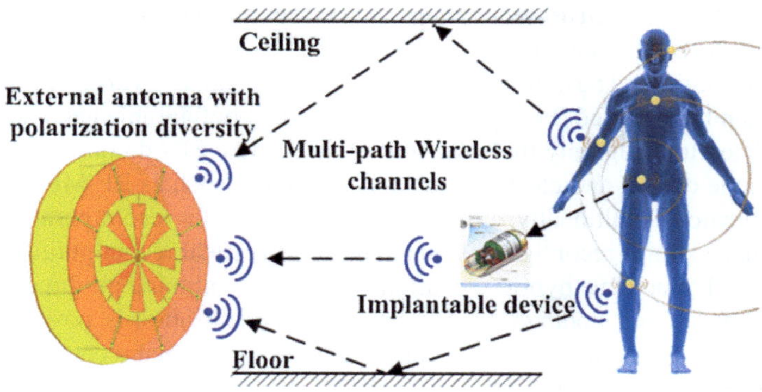

Fig. 9.3 Body-centric wireless communication system [3]

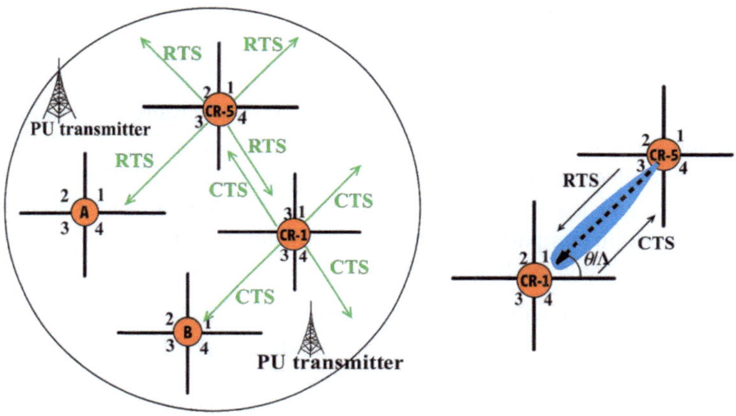

Fig. 9.4 An example of using hybrid reconfigurable (frequency-pattern) antenna in a cognitive radio network [9]

overcome the problem, a hybrid reconfigurable antenna made by combining frequency and radiation pattern reconfigurability will be best suited. Here, the omnidirectional radiation mode is preferred to transmit the control signals, such as a request to send (RTS) and clear to send (CTS). On the other hand, the data transmission should be done in directional mode. Therefore, a frequency/pattern hybrid reconfigurable antenna will be best to use. Since control signals like RTS and CTS are shared regularly to check and maintain the connectivity among devices, omnidirectional transfer is desired. For data transfer, if the position of the receiver is known, then directional data transfer will save a lot of radiation power. A tentative network configuration with the primary user transmitter and cognitive radio (CR) nodes with control packets such as RTS, CTS, and angle of arrival (AoA) is mentioned in Fig. 9.4 [9].

PIN diodes are used to reconfigure antenna frequency. If the same antenna's corners are truncated, then it will achieve properties that switch from linear to circular polarization. Thus, a frequency/polarization-based hybrid reconfigurable antenna can be designed. Furthermore, the hybrid reconfigurable antenna design is not restricted to using only the solid-state switches like PIN diode, varactor diode, and FET; the optical and electro-mechanical switches such as RF-MEMS can also be implemented. Additionally, the smart materials can offer reconfigurability and be used to make hybrid reconfigurable antennas. Geometrical reconfigurability can be implemented to achieve hybrid reconfigurability by combining all three types: frequency, pattern, and polarization. This type of hybridization is possible but can cause complexity and unanticipated consequences. An optimal reconfigurable technique depends on the application.

9.2.2 *Mechanisms for Implementing Antenna Reconfigurability.*

Based on the switching mechanism, antenna reconfigurability is broadly categorized among solid-state, electro-mechanical, optical, and smart material switches. Solid-state and electro-mechanical are more popular and referred as conventional switching mechanisms in reconfigurable antennas. Optical and smart material switches are used in comparatively modern applications. The details of the different types of switches are given below [10].

(a) **Solid-State Switches:** Electronic switches such as PIN diodes, varactor diodes, and FET can be used to achieve reconfigurability. The PIN diode is popularly used as it offers very low insertion losses. The PIN diodes are applicable for reconfigurability across the whole RF range while varactor diodes have a unique feature: the flow of current is controlled. The blocking capacitor is not required in the FET implementation to achieve reconfigurability. Most electronic switches are fast but cause losses due to nonlinearities [11].

(b) **Electro-mechanical Switches:** Mechanical actuators are used to provide the movement, mainly rotations. High efficiency can be achieved at the cost of durability. The movement through actuators is slow. An increase in switching speed is desired to make the switches more effective. The electro-mechanical switches such as RF-MEMS and RF-NEMS are used for this purpose. The issues of non-linearity in PIN and varactor diode are overcome in the RF-MEMS switches. Therefore, the noise in the system is less. Still switching speed in electro-mechanical switches will remain comparatively less than the solid-state switches. The higher voltage requirement is another disadvantage of such devices [12].

(c) **Optical Switches:** Photodiodes are mostly used as optical switches. The main advantage of the optical switches is their frequency response and can operate at effectively at very high frequencies. The instability caused by temperature rise is the major challenge in the designing of reconfigurable devices with photodiodes.

(d) **Smart Material Switches:** Material-based reconfiguration can be implemented through the ferrite and liquid crystals. The tuning of such devices is very smooth, but control is complex. The photoconductive materials are also used to alter the antenna properties rapidly. The power requirement is lower for such devices. However, achieving reconfigurability with such devices is comparatively costly.

The details of the advantages, disadvantages, and limitations of different reconfigurable mechanisms with switches are given in Table 9.3.

Table 9.3 Classification of reconfigurable switches, advantages, and disadvantages

Reconfigurability technique	Switch type	Advantages	Limitations/ disadvantages
Solid-state switches	PIN diode	1 Low cost and easy to attach and fabricate 2 Low insertion losses 3 Reliable in switching 4: Useful in 10 MHz to 100GHz range	1 Very high DC biasing is required 2 Blocking capacitor limits the performance 3 High tuning speed causes synchronization problems
	Varactor diode	1 Current flow can be controlled. 2 Easy in antenna integration 3 Sustained tuning	1 Dynamic range is limited 2 Non-linearity of diode causes additional noise and losses
	Field effect transistor (FET)	1 Biasing is easy because the blocking capacitor is not required 2 Switching speed is high 3 Cost efficient	1 Performance suffers at high frequencies as poor isolation is offered 2 Highly non-linear and causes insertion losses
Electro-mechanical switches	RF-MEMS/ NEMS switch	1 Offers linearity and, therefore, low noise figure and excellent isolation 2 Low-power consumption	1 Switching speed is poor 2 Cost is high 3 The actuator requires high voltages
Optical switches	Photo diodes	1 Offers linearity 2 Excellent frequency response 3 Offers low resistance	1 Rise in temperature causes instability 2 Poor sensitivity 3 An offset voltage is required for operation
Smart material switches	Ferrite and liquid crystal material	1 Power consumption is less 2 Geometric distortions are suppressed 3 Zero emission of radiation	1 Response time is poor 2 Expensive 3 Radiation angles are restricted

The key features that made reconfigurable antenna so beneficial in adaptive healthcare are as follows:

1. The reconfigurable antennas are adaptive to adjust their parameters according to the surrounding environment.
2. The body movement and postures cannot affect the communication with the outdoor monitoring system. The antenna is adaptive and can adjust its operation according to the requirement.
3. Inside the human body, this adaptability helps to sustain communication irrespective of changes in tissue size, accumulation of fats, variation of the concentration of blood, or other physiological changes.
4. A single reconfigurable antenna can sufficiently replace multiple fixed antennas. Therefore, the design of the healthcare device can become very compact.
5. In healthcare applications, the reconfigurable antenna can be used at the transmitter, i.e., at the patient bedside unit in telemetry communication, as a wearable antenna or implant antenna, or as a receiver at the base station unit.
6. The reconfigurable antenna makes healthcare devices energy efficient. The antenna achieves energy efficiency by implementing different combinations of ON and OFF of the diode switches during operation. The feature is particularly useful in implantable and wearable devices.
7. Because of reconfigurability, the healthcare devices can be customized to different wireless standards. FCC has provided many standard standards for healthcare communication, followed by reconfigurable antennas. This becomes possible because the reconfigurable devices can support different protocols at the same time.
8. The beamforming can also be implemented with reconfigurable antennas for uninterrupted off-body communication.
9. The reconfigurable antenna arrays are used in some applications to achieve a higher degree of directivity. Off-body telemetry communication, such as an ambulance to the hospital, requires an antenna array.

9.3 Applications in Adaptive Healthcare Devices

1. **Wearable Devices:** Because of dynamic frequency adjustment, adaptive healthcare devices can be suited to wearable technology for real-time monitoring. The device will remain functioning even in very high-frequency activities. Remote monitoring is gaining popularity with advancements in technology related to the Internet of Medical Things (IoMT). The sensors are knitted or weaved into the clothes and thus placed near the patients to monitor diverse continuous physiological parameters. The design of the wearable device is dependent on the patient's body composition and the location of the antenna, i.e., the antenna is placed near which part of the body. In addition, the movement of that particular body part. An antenna with an adaptive feature will be best suited to ensure the

best communication link. The reconfigurable antenna provides the advantages to mitigate the said challenges [13, 14].

2. **Implantable Devices:** The adaptive devices can be used as implantable devices to maintain communication with multiple off-body base stations. The key parameters that can affect the implant device's performance are the thickness, dialectic properties, conductivity, water content, and density of the tissue. The antenna's effectiveness is also dependent on the depth of the antenna placement from the skin. The pattern reconfigurable antenna is preferred to steer the beam from dense tissues so more radiation can come out of the body. The pattern reconfigurability of adaptive devices can be very useful in addressing the SAR regulations for a tissue [15] (Fig. 9.5).
3. **Telemedicine:** Seamless communication with distant professionals is the need of the hour. The adaptive devices will help healthcare professionals connect to the patient's monitoring system across different networks and geographies. Therefore, a reliable communication link is set up between the transmitter and the receiver. The main advantage of telemedicine technology is that it can reduce the burden on health professionals to monitor individual patients at distant locations. The health official can monitor multiple patients simultaneously from one location. The use of reconfigurable antennas can help achieve this by tuning the different frequencies for different patients [17].

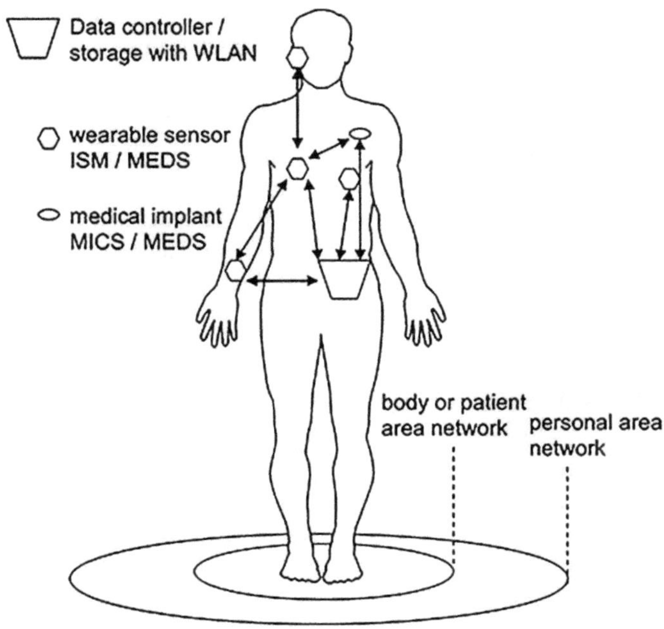

Fig. 9.5 Body area devices and network [16]

Consider an example of an intensive care unit in a hospital where officials have to monitor multiple patients simultaneously. The radiation pattern reconfigurable antenna installed at the patient's bed unit will direct the radiating beam to the base station from all directions. In this example, the pattern reconfigurable antennas are installed at the patient's bed units, and at the receiver, the omnidirectional antenna is installed. The frequency reconfigurable antenna is also useful in the same scenario and is installed at the receiver base station side, where the different patient's bed units will transmit at different frequencies.

4. **Internet of Medical Things (IoMT):** The sensors are an integral part of IoMT devices, and many of these devices operate at the same frequency. While different devices use different communication protocols and modulation schemes, in the channel, the frequency will be the same. Consider an example of ISM band 2.4 GHz, the frequency is unlicensed so any user can utilize the frequencies. On the other hand at 2.4 GHz, the communication standards such as Bluetooth and Wi-Fi also work. Although the devices use different communication protocols, but the channel is same hence suffer with the interference. Integrating reconfigurable antennas with these devices can allow simultaneous data collection from different IoMT devices as they can alter their transmission frequency. In the figure given below, an IoMT model is shown. The patient is monitored by different sensors, either implanted or wearable. The data from patients can reach the IoT cloud and then go for processing. Sometimes, the data can be directly monitored by doctors or health officials, and corresponding treatment is advised to the patients. Since the data is coming from different source sensors so the type and format of data cannot be the same. This is also shown in the figure as the sensors measure the heart rate and blood pressure. Since these two sensors work on different principles so their data type cannot be the same. Therefore, a separate set of analyzers and processors is necessary to give a concrete outcome [18].

Here, the reconfigurable antenna is applicable for installation with sensors for data transmission. In such cases, both the heart rate monitor sensing unit and the blood pressure sensing unit employ reconfigurable antennas. At one frequency, the data is sensed by the heart rate sensor, and at another frequency, data from the blood pressure sensor is transmitted to the IoT cloud (Fig. 9.6).

9.4 Frequency Range: Application-Specific

Adaptive healthcare devices are gaining popularity nowadays. The capability to tune multiple frequencies is a key feature that enables many applications for these devices. Therefore, the frequency reconfigurable antenna as a key component of adaptive dives can help to achieve frequency tuning. There are a few frequency ranges commonly employed in medical applications. The broader categorization of these applications is done among body area network communication, diagnostic and imaging systems, wireless power transfer, Internet of Medical Things (IoMT)

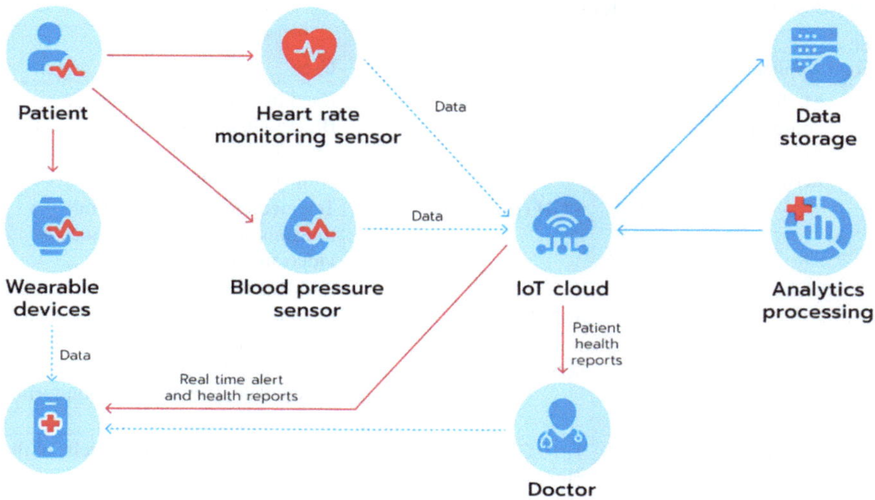

Fig. 9.6 Internet of Medical Things (IoMT) for real-time medical examination [19]

devices, non-invasive heating for therapeutic applications, and biosensing. Table 9.4 gives the application-specific frequency ranges and their uses [20–23].

9.5 Design Parameter and Specification

The following specifications play an essential role in designing an adaptive healthcare device. The reconfigurable antennas can be used in implantable, wearable, and telemedicine devices. And their design specification varies accordingly.

(a) **Compact size:** Size is the most crucial factor for implantable and wearable medical devices. The reconfigurable antenna used in wearable devices should be compatible with body movement and posture. The challenge is big in the case of implanted antennas as they should come inside human tissue.
(b) **Frequency ranges:** Another crucial design specification is the frequency of operation. The antenna should have a specific frequency range, as mentioned in the table above. For example, for the MICS application, the antenna should have resonance in the 402–406 MHz range.
(c) **Biocompatibility of the device:** In the case of an implantable reconfigurable antenna, the biocompatibility of the substrate material is an important parameter. So, the antenna will have safe integration with the implanted tissue. The flexible materials such as liquid crystal polymer can be used to achieve biocompatibility.
(d) **Reconfigurability:** The frequency reconfigurable antenna should operate in multiple frequency bands to satisfy the different applications. Further, an elec-

Table 9.4 Frequency bands for medical and their uses

Sr. no.	Applications	Frequency ranges	Uses
1	Medical device communication (wireless body area network-WBAN)	2.4 GHz (ISM band)	Used in wearable technology, Bluetooth, and Wi-fi communication. Real-time monitoring of heart rate, temperature, etc.
		5 GH (ISM band)	Telemedicine—High data rate transmission applications such as video streaming to distant professionals
		402–405 MHz (medical Implant communication service-MICS)	Seamless communication between implant and off-body base station. Examples of implant devices—Pacemakers and glucose-monitoring devices
2	Diagnostic imaging systems	1–10 GHz (microwave range)	Early-detection of cancer cells in the breast and brain
		400 MHz—3GHz	The RF energy is used for ablation therapies to destroy abnormal tissue and tumors
		Below to 27 MHz	In magnetic resonance imaging (MRI) systems
3	Wireless power transfer (WPT)	13.56 MHz (NFC)	Power transfer to near-field implant devices
		6.78 MHz (industrial band)	To enable wireless charging of implant devices
		915 MHz (ISM band)	Used for inductive power transfer to remote devices
4	IoMT (internet of medical things)	433 MHz, 868 MHz and 915 MHz (sub 1GHz band)	Remote health monitoring through long-range communication
		2.4 GHz (ISM band)	Provide connectivity to many high data rate networks. Cloud communication
		NR band (24.25–27.5 GHz)	Low latency and very high reliability for health monitoring
5	Therapeutic methods	27.12 MHz	Diathermy devices are used to heat up the tissue for physiotherapy
		2.45 GHz	Microwave range for hyperthermia in cancer treatment
		3–30 Mhz	In the electrotherapy of muscles
6	Biosensing	30–300 GHz (mm-wave)	Advanced diagnostic: Detection of biochemical markers
		0.1–10 THz	Non-invasive blood glucose monitoring

tronic switching mechanism must be implemented to alter the operating bands according to our requirements. In the case of the pattern reconfigurable antenna, the switches must be configured to different modes to steer the radiation beam according to the need. The modes are configured to change the type of polarization in the case of polarization reconfigurable antennas. Therefore, selecting different operational modes is crucial in designing the reconfigurable antenna.

(e) **Energy efficiency:** In the implantable antenna, the energy efficiency is a crucial parameter as the antenna is battery power driven. And it is tough to change the battery frequently. In wearable antennas, energy efficiency is required as frequent knitting or weaving of the devices is not possible.

9.6 Problem Formulation

Wireless medical applications working near 1.3 GHz frequency are typically used in medical imaging and body area networks. In the field of diagnosis and imaging, the 1.3 GHz frequency is used for early detection of breast cancer and brain stroke diagnosis as it provides a balance between penetration depth and resolution. This frequency range shows non-ionizing effects and is, therefore, safe for human tissue. In the case of body area networks, the 1.3 GHz frequency range offers moderate range and penetration through tissues with low latency. Since the sensor is placed on or in the body, the frequency range benefits in the transmission of data wirelessly. Therefore, the frequency range is ideal for real-time vital monitoring.

From the above discussion, it is clear that the body area network is the prominent application of the frequency band near 1.3 GHz. Further, the 1.3 GHz band is not part of the ISM band. This means that the 1.3GHz band is a licensed band, and the users have to buy spectrum to operate the band. Since the transmission range of the WBAN at 1.3 GHZ frequency becomes limited, then it is advantageous to use an ISM band at 2.4 GHz. The ISM band provides global coverage, interoperability, and compatibility to merge with other technologies. These features made the 2.4 GHz ISM band one of the prominent bands to use in the unlicensed space.

Considering all the features, problems for the current chapter are formulated as follows:

- A frequency reconfigurable antenna is designed to operate in 1.3 GHz and 2.4 GHz frequency bands.
- The antenna should operate at a 1.3 GHz band to be suitable for wireless body area network (WBAN) applications.
- The 1.3 GHz band provides moderate performance in terms of range and penetration.
- The reconfigurability in the antenna is achieved by implementing a PIN diode.
- The 2.4 GHz frequency band has applications in wireless medical telemetry communication. Can be used for medical.

9.7 Related Work

Researchers have extensively explored the reconfigurable antenna design for a variety of applications such as implant, wearable, body area network, and wireless medical telemetry. Different types of reconfigurability, such as frequency, radiation pattern, polarization, and hybrid explored by the researchers. Key studies include the literature survey mentioned in Table 9.5.

9.8 Simulation Results and Discussion

The proposed frequency reconfigurable antenna has been designed for biomedical applications. The antenna was designed on an FR4 substrate with a thickness of 1.6 mm. The antenna covers the full ground plane, as shown in Fig. 9.7. The dimensions of the proposed reconfigurable antenna are shown in Table 9.6.

The reconfigurable antenna operates in two states. One state when the diode is ON, and a second state when the diode is OFF, the antenna resonates at 2.45 GHz and covers the bandwidth from 1.87 GHz to 6.52 GHz. The resonance frequency and its bandwidth are shown in Fig. 9.7. In the second state, when the diode is ON, the antenna resonates at two frequencies, 1.43 GHz and 2.68 GHz resonance frequencies, and at these resonance frequencies, the antenna covers a bandwidth of 1.25 GHz to 1.73 GHz and 2.2 GHz to 6.58 GHz, respectively. The second state resonance frequencies and bandwidth are shown in Fig. 9.8.

The resonance behaviors of the antenna can be further understood with the help of the current distribution diagram. In the first state, when the diode is OFF, most of the current flows in the rectangular patch, as shown in Fig. 9.9. In the second state, when the diode is ON, the current flows in all the antennae, as shown in Fig. 9.10.

The radiation performance of the antenna can be understood by 2D radiation (E plane and H plane). In state first, when the diode is OFF, the antenna resonates at 2.45 GHz. The radiation performance of the antenna at 2.45 GHz is shown in Fig. 9.11. This antenna provides a bi-directional radiation pattern in the E plane and an omnidirectional radiation pattern in the H plane. In state second, when the diode is ON, the antenna resonates at 1.43 GHz and 2.68 GHz resonance frequencies. The radiation pattern (E plane and H plane) at both frequencies is shown in Figs. 9.12, 9.13 and 9.14.

9.9 Open Research Challenges

Designing reconfigurable antennas for adaptive healthcare devices is popular nowadays. Though several advancements have been made in reconfigurable antenna design techniques for adaptive healthcare devices, a few challenges have still not been tackled or have scope for improvement.

Table 9.5 Literature review

Paper	Reconfigurability type	Key feature	Details
[6]	Frequency reconfigurable	Antenna design based on split-ring resonators	In the proposed paper, a switch is used between the split-ring elements. A probe feeding method has been employed. The overall size of the antenna is $12 \times 12 \times 0.635$ mm^3. In this paper, the antenna performance is observed in terms of S11 characteristics, radiation patterns, and surface current distributions
[7]	Radiation pattern reconfigurable	Metamaterial structure in wearable antenna application	The antenna resonates at the 0th order mode or the + 1 mode and yields a broadside or an omnidirectional radiation pattern, respectively. The textile material is used in the fabrication of the antenna. It covers the ISM band at around 2.4 GHz. For analyzing the performance, the antenna is also simulated on a human tissue model. The obtained SAR level is well below the European standard threshold
[8]	Polarization reconfigurable	Textile patch antenna for wearable IoT applications	PIN diodes are employed as switches on the truncations of patch antenna to achieve polarization reconfigurability. The antenna shows a resonance in in ISM band near 2.4 GHz with an average gain of 5.96 dBi among different polarizations. The elements of the antenna are designed with a fabric material so that the antenna can integrated with the wearable cloths. A body-mimicking phantom is used for antenna performance analysis
[10]	Hybrid reconfigurable (frequency-pattern reconfigurable)	A center-shorted microstrip patch antenna	The antenna design achieves two resonance modes with shorting at the center of the patch. Two groups of varactor diodes are used with tuning range of [0.149, 1.304] pF. The diodes are placed at opposite sides of the antenna. Also, open-circuited loading stubs are used. The antenna structure radiates as a dual-band antenna with broadside radiation at the higher resonance frequency and monopole radiation at the lower resonance frequency. By changing the bias voltage, these resonance frequencies can be simultaneously moved. The proposed antenna offers more than 20% continuous fractional frequency range

(continued)

Table 9.5 (continued)

Paper	Reconfigurability type	Key feature	Details
[13]	Frequency reconfigurable	Wearable antenna for 5G health monitoring applications	The wearable antenna is designed on a transdermal cotton patch substrate. PIN diodes are used to achieve reconfigurability. The achieved antenna shows resonance at three different frequencies of 1.8 GHz, 2.3 GHz, and 3.5 GHz. The frequencies are part of the 5G range in Malaysia and Indonesia. Further, antennas are tested in different bending conditions at all the resonating frequencies. The obtained results seem satisfactory
[14]	Pattern reconfigurable	Textile antenna array for wearable applications	In the paper, the fluidically reconfigurable antenna is studied and reported. A loop antenna is designed. A four-channel platform is placed on top of the antenna for beam steering. The distilled water and seawater are used as fluidics in actuators. Each channel is activated by one fluid. The simulation result shows that at the frequency 1 GHz, and with the use of fluidics distilled water and seawater, the elevation angle can be steered between -145° and 139°. In the respective configuration, the realized gain amplitude is varied between 1 and 15 dB
[15]	Frequency reconfigurable	MIMO antenna for injectable in-body implant Applications	The miniaturization is achieved using meandered lines and shorting pins, while reconfigurability is obtained by using the diode as a switch and by adjusting the meandered length. The antenna resonates between 2.42–2.48 GHz and 2.38–2.48 GHz. The antenna is fabricated on a flexible polyimide substrate ($\varepsilon r = 3.5$), with dimensions of $5 \times 2 \times 0.05$ mm^3. The performances are evaluated through single-layer (fat/muscle) and three-layer (skin, fat, and muscle) phantoms. The tissue-independent MIMO antenna achieves isolation higher than 15 dB, with impedance bandwidths of 60 MHz in fat and 100 MHz in muscle
[23]	Polarization reconfigurable	PIFA for implantable telemetry Applications	Two groups of PIN diodes are used as RF switches along the x and y directions. The two orthogonal linear polarizations are archived. The proposed PIFA can be used as an implanted antenna. For validation of the designed antenna, an in vitro test was carried out in body tissue. The results show an impedance bandwidth of 9.1% from 2.3 GHz to 2.52 GHz in different reconfigurable states. The radiation performances, SAR, and link characteristics are also tested in the paper

(continued)

Table 9.5 (continued)

Paper	Reconfigurability type	Key feature	Details
[25]	----	A multiband antenna for biomedical telemetry applications	The tri-band antenna resonates at 375 MHz, 2.525 GHz, and 5.625 GHz resonance frequencies. The antenna is designed on flexible Rogers RT/droid 5880 epoxy substrate and covers the size of $70.0 \times 80.0 \times 1.0$ mm^3
[26]	Frequency reconfigurable	Designed a reconfigurable planner antenna for neural network applications	The antenna is designed on a met surface superstrate. The antenna is designed on a dual-layer 1.524 mm thick Rogers RO4350B substrate. The antenna is designed and simulated using HFSS software. The frequency reconfigurability of the antenna is achieved by metasurface superstrate on the patch antenna. The antenna reconfigured from the frequency range from 5.03 to 6.13GHz
[27]		Reconfigurable antenna based on the metamaterial for wearable applications	The antenna is designed with a monopole and Chebyshev transformer coupled with a unit cell of fractal Minkowski curve. The antenna is designed on an FR4 substrate with a 1.6 mm thickness. The antenna is designed on a single-side metal plane on the substrate. The antenna performance without EBG resonates at 2.45 and 2.7 GHz, which provides the gain (efficiency) of 2.74 dBi (80.6%) and 2.61 dBi (95.5%), respectively. The antenna with EBG resonates at 2.4 GHz, 2.7 GHz, and 3.4 GHz provides the gain (efficiency) of 4.93 dBi (93%), 5.0 dBi (93.8%), and 1.5 dBi (60.3%), respectively
[28]		X-shape slotted patch antenna for implementable device applications	The antenna is designed by cutting a slot from a radiating patch with a co-axial fed. The antenna is designed on a flexible Roger Duroid RT5880 substrate with a thickness of 0.254 mm substrate. The antenna covers the bandwidth from 800 MHz to 1.0 GHz inside the tissue. The SAR of the antenna was achieved at 8.17 W/kg over 1 g of mass tissue
[29]		Circular patch antenna for biomedical applications for 5G at frequency 2.45 GHz	The flexible polymer-based antenna is designed for 2.4 GHz frequency. The radiating element is a circular patch. The antenna resonates at 2.45 GHz and covers 140 MHz bandwidth. The antenna provides 95% efficiency with 5.8 dB gain
[30]		Reconfigurable spiral-shaped monopole antenna for biomedical applications	The switching operation of the antenna is achieved by using microwave switches. The antenna is designed on an FR4 substrate and is capable of operating at 402 MHz to 406 MHz and 2.4 GHz to 2.5 GHz ISM bands. The designed antenna covers the size $32.0 \times 50.3 \times 1.8$ mm^3

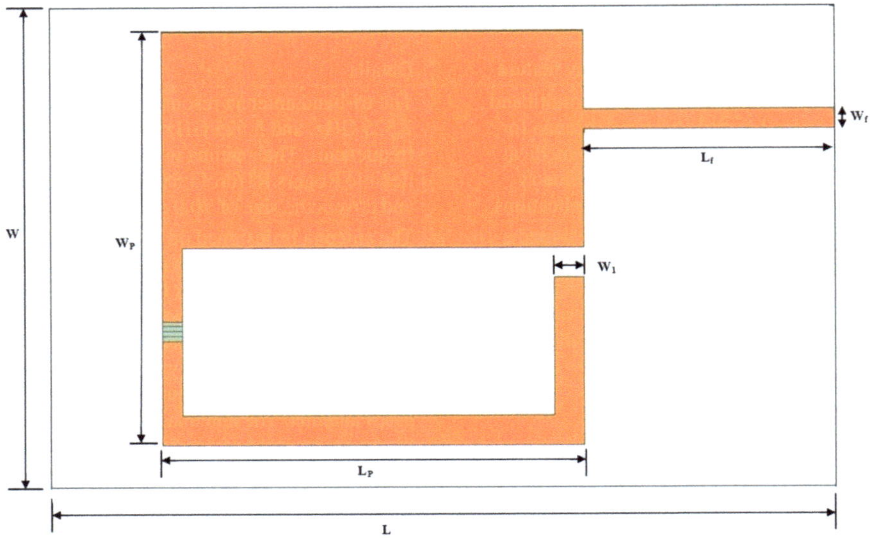

Fig. 9.7 Proposed frequency reconfigurable antenna

Table 9.6 The dimensions of the proposed frequency reconfigurable antenna

Parameters	Dimensions (mm)
L_p	21.0
W_1	1.5
L	39.0
L_f	12.5
W_f	1.0
W	24.0

- Power management is necessary to ensure a longer operational life for implantable and wearable reconfigurable antennas. These antennas should consume low power to enlarge the battery life.
- With the innovation coming in communication technology like 5G and 6G, new standards are being developed. The reconfigurable antennas should be compatible with these new standards.
- Further, new communication technologies offer ultra-high throughput and very low latency. To comply with these features, the reconfigurable antenna should fall within the communication range of these technologies.
- Innovations are happening to develop new biocompatible materials. The reconfigurable antenna should be developed with these materials to check their response in human tissue.

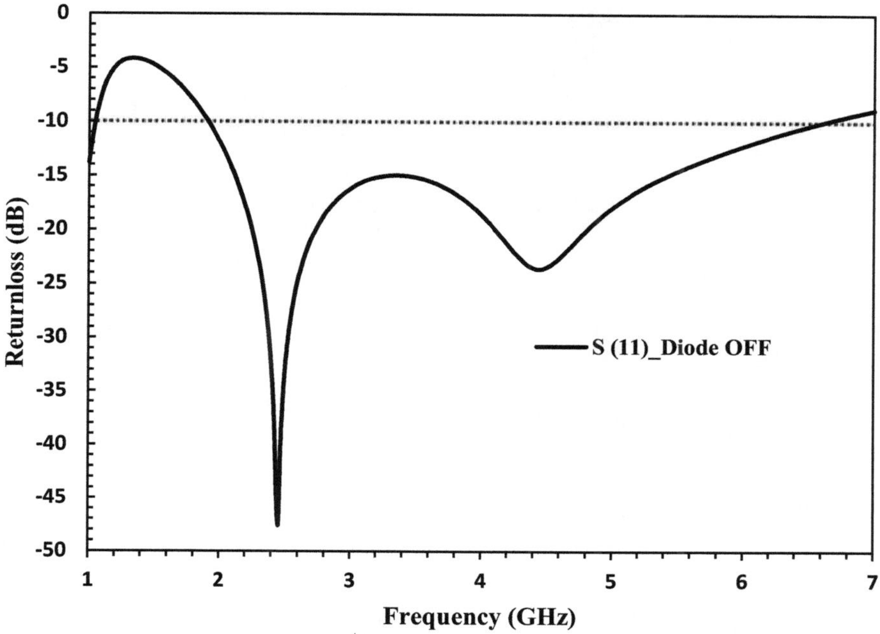

Fig. 9.8 Return loss of the proposed antenna when the diode is OFF

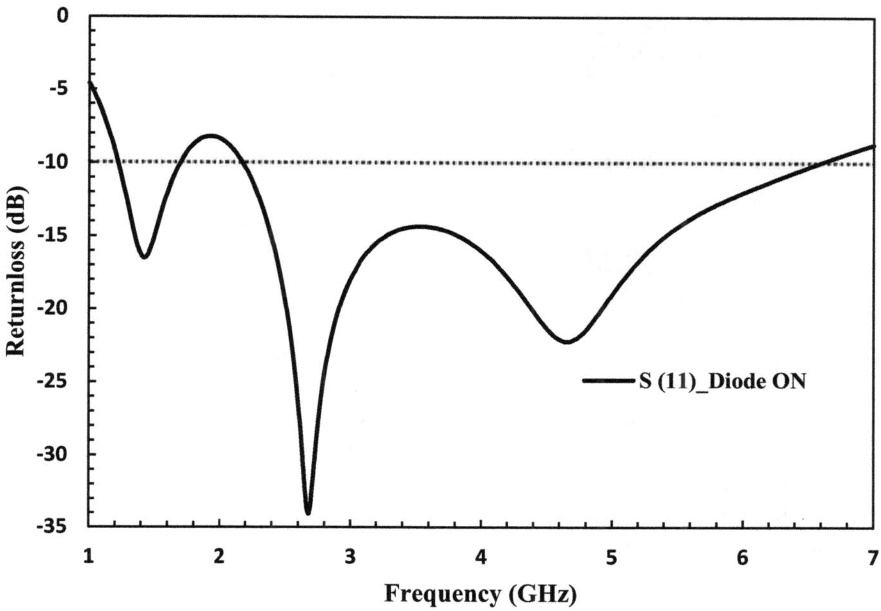

Fig. 9.9 Return loss of the proposed antenna when the diode is ON

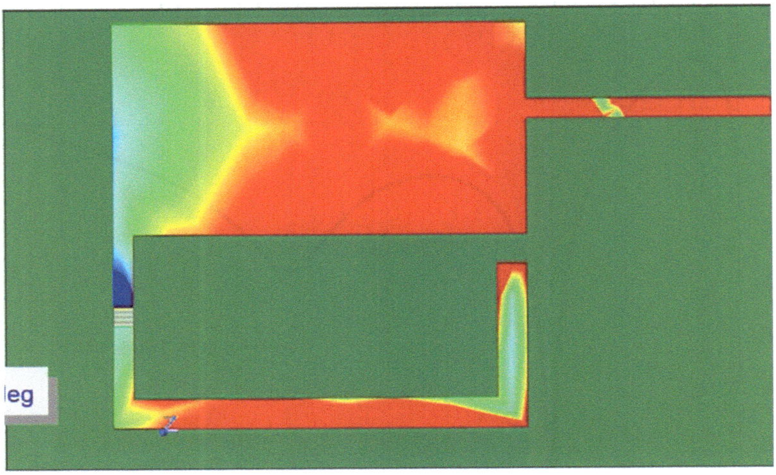

Fig. 9.10 Current distribution of the proposed antenna when diode is OFF

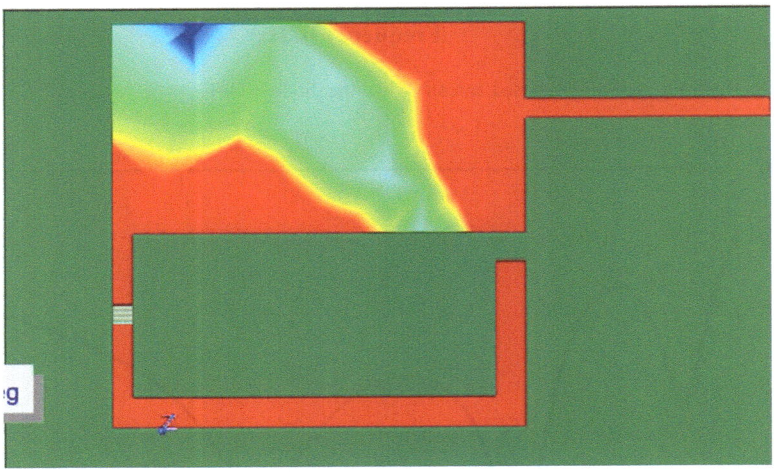

Fig. 9.11 The current distribution of the proposed antenna when the diode is ON

- Miniaturization is needed for many adaptive healthcare devices. However, a trade has to be made between the performance of the devices and miniaturization.
- Carrying out comprehensive testing for the adaptive devices is still a challenge that has to be addressed.

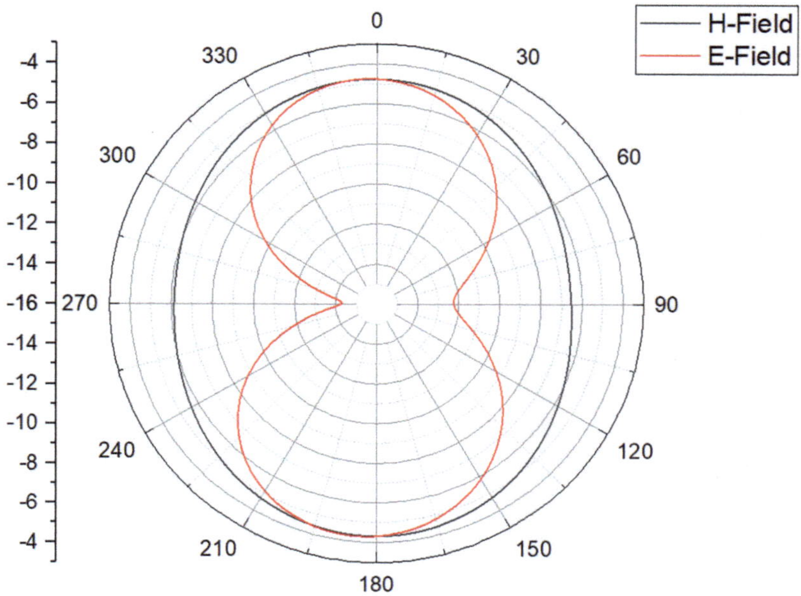

Fig. 9.12 The 2D radiation pattern of the antenna at 2.45 GHz

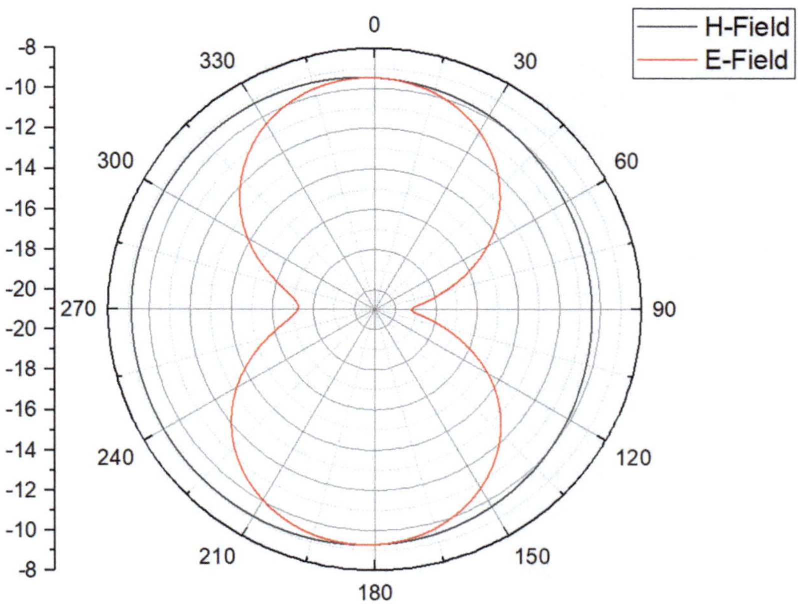

Fig. 9.13 2D radiation pattern of the antenna at 1.43 GHz

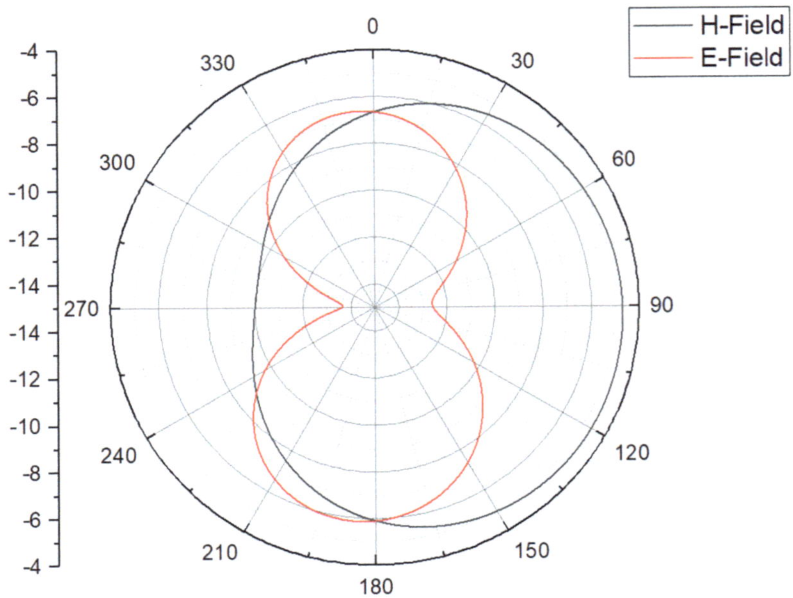

Fig. 9.14 2D radiation pattern of the antenna at 2.68 GHz

9.10 Summary

The reconfigurable antennas in adaptive healthcare devices offer new modalities to users to change the frequency bands, radiation patterns, and polarization. The flexibility of parameter variation to tune to frequency, pattern, or polarization makes the antenna versatile. The antenna becomes compatible with the latest wireless communication standards offered by modern technologies. The chapter starts with an introduction where key concepts of reconfigurability are explained along with the definitions and uses. Here, the types of reconfigurable antennas and the different mechanisms to achieve reconfigurability are discussed in detail. The next topic covers applications of reconfigurable antenna in adaptive healthcare. Further, the frequency ranges used for a reconfigurable antenna in adaptive healthcare applications are covered. Next, the different design parameters for the reconfigurable antenna design are covered. After this, the problem of a frequency reconfigurable antenna design is formulated. Here, a frequency reconfigurable antenna is designed, and a PIN diode is used as a switch. The two bands, 1.3 GHz and 2.4 GHz, are covered, which make the antenna applicable in WBAN and medical telemetry applications. Further the latest research work is surveyed in the topic literature survey. Then, the simulated results of the antennas are obtained and discussed in detail. Finally, the open research challenges in the field are covered in the chapter.

The reconfigurable devices achieve ultra-high throughput and low latency. However, the reconfigurability of antennas used in medical applications offers many advantages, the technology still has some challenges that need to be addressed. Power management, compatibility with new technology and standards, miniaturization, biocompatible materials, and testing of devices in the real world are a few of them. Further, the dependency of physiological monitoring, contactless therapies, and biosensing on wearable and implantable devices is expected to grow in the future. The use of reconfigurable antennas in wearable and implant technology must be fulfilled to optimize design and fabrication. Flexible material selection, linear characteristics of the switching device, SAR limit, reconfiguration mechanism, and bending investigation are the key characteristics that need to be observed while implementing reconfigurability in wearable and implant antennas. The use of the material is very crucial in implant and wearable applications. The dielectric constant governs the electrical properties of the antenna, and along with the thickness of the substrate, the linearity property of the antenna is defined. Recently, metallic ink, conductive polymers, and Zelt have gained popularity as conductive materials for effective EM radiation. To observe the effect of the radiation on the human body, FCC and ICNIRP have provided the regulations for the SAR limits. Not only the absorption but other issues of the tissues have to be taken care of, such as moisture, temperature, and durability. In the case of an implant, the reconfigurable antenna has to be bent or conformed to a certain degree inside the body so that it can come out effectively. Hence, investigation of different bending conditions is necessary. To design a hybrid reconfigurable antenna for a particular application, a comparative performance analysis among different switching mechanisms, such as PIN diode, FET, varactor diode, and RF-MEMS, has to be conducted.

From the above discussion, we can conclude that the full potential of reconfigurable antennas in adaptive healthcare has not been explored yet. Research to resolve the existing challenges needs to be conducted in the future.

References

1. E. García, A. Andújar, J. Anguera, Overview of reconfigurable antenna systems for IoT devices. Electronics **13**, 3988 (2024). https://doi.org/10.3390/electronics13203988
2. O.M. Khan, K. Raza, R. Shubair, Q.U. Islam, Frequency reconfigurable implant antenna for MICS and ISM band applications, in *2018 18th International Symposium on Antenna Technology and Applied Electromagnetics (ANTEM)*, vol. 2018, (Waterloo, ON, Canada), pp. 1–2. https://doi.org/10.1109/ANTEM.2018.8572932
3. L. Huitema, H. Wong, W. Lin, A. Crunteanu, Reconfigurable multi-slot antenna for biomedical applications, in *2016 10th European Conference on Antennas and Propagation (EuCAP), Davos*, (2016), pp. 1–4. https://doi.org/10.1109/EuCAP.2016.7481209
4. J.K. Rai, P. Ranjan, R. Chowdhury, Recent development in reconfigurable dielectric resonator antenna and microwave filter: design and application. Int. J. Commun. Syst. **37**(14), e5859 (2024). https://doi.org/10.1002/dac.5859
5. L.A. Selsiya, R. Janani, J. Gopalakrishnan, K. Jayamani, Low profile frequency reconfigurable antenna for ISM band applications, in *2021 International Conference on System, Computation, Automation and Networking (ICSCAN), Puducherry, India*, vol. 2021, pp. 1–4. https://doi.org/10.1109/ICSCAN53069.2021.9526419

6. M. Ciflik, S.C. Basaran, Frequency reconfigurable antenna Design for Tissue Independent Medical Applications at 2.4 GHz ISM band, in *2023 14th International Conference on Electrical and Electronics Engineering (ELECO), Bursa, Turkiye*, vol. 2023, pp. 1–3. https://doi.org/10.1109/ELECO60389.2023.10416027
7. S. Yan, G.A.E. Vandenbosch, Radiation pattern-reconfigurable wearable antenna based on metamaterial structure. IEEE Antennas Wirel. Propag. Lett. **15**, 1715–1718 (2016). https://doi.org/10.1109/LAWP.2016.2528299
8. H. Lee, J. Choi, A polarization reconfigurable textile patch antenna for wearable IoT applications, in *2017 International Symposium on Antennas and Propagation (ISAP), Phuket, Thailand*, vol. 2017, pp. 1–2. https://doi.org/10.1109/ISANP.2017.8228882
9. S. Anamalamudi, M. Jin, Energy-efficient hybrid CCC-based MAC protocol for cognitive radio ad hoc networks. IEEE Syst. J. **10**, 358–369 (2016). https://doi.org/10.1109/JSYST.2014.2367504
10. N. Nguyen-Trong, L. Hall, C. Fumeaux, A frequency- and pattern-reconfigurable center-shorted microstrip antenna. IEEE Antennas Wirel. Propag. Lett. **15**, 1955–1958 (2016). https://doi.org/10.1109/LAWP.2016.2544943
11. P.K. Devi, M. Sujatha, J.A. Prasath, Design of reconfigurable integrated patch antenna in ISM band for IoT applications, in *2020 7th International Conference on Smart Structures and Systems (ICSSS), Chennai, India*, (2020), pp. 1–4. https://doi.org/10.1109/ICSSS49621.2020.9202152
12. S. Majumder, J. Lampen, R. Morrison, J. Maciel, MEMS switches. IEEE Instrum. Meas. Mag. **6**(1), 12–15 (2003). https://doi.org/10.1109/MIM.2003.1184267
13. L. Hendrajaya, B. Babu, N.A. Samsuri, B.S. Nugroho, M.K.A. Rahim, The performance of a frequency reconfigurable wearable antenna for 5G health monitoring application under bending conditions, in *2023 IEEE International Symposium on Antennas and Propagation (ISAP), Kuala Lumpur, Malaysia*, vol. 2023, pp. 1–2. https://doi.org/10.1109/ISAP57493.2023.10388638
14. D. Vital, A novel fluid-based pattern-reconfigurable, textile antenna Array for wearable applications, in *2022 IEEE International Symposium on Antennas and Propagation and USNC-URSI Radio Science Meeting (AP-S/URSI), Denver, CO, USA*, vol. 2022, pp. 1028–1029. https://doi.org/10.1109/AP-S/USNC-URSI47032.2022.9886570
15. L. Harlan, M. Susila, S. Kumar, H. Chul Choi, K. Wook Kim, Reconfigurable ultra-miniaturized MIMO antenna for tissue-independent communication in injectable medical implants. IEEE Access **13**, 22663–22677 (2025). https://doi.org/10.1109/ACCESS.2025.3535780
16. P.S. Hall, Y. Hao, *Antennas and propagation for body-centric wireless communications* (Artech House, MA, 2006)
17. N. Haritha, E. Gowsika, P. Elangovan, K. Jayamani, Design of Wearable Reconfigurable Antenna for IoT applications, in *2021 International Conference on Advancements in Electrical, Electronics, Communication, Computing and Automation (ICAECA), Coimbatore, India*, vol. 2021, pp. 1–5. https://doi.org/10.1109/ICAECA52838.2021.9675735
18. M.M. Farhad, A. Jayed Islam, S. Chakraborty, M.S. Bin Nesar, M.A. Siddique, N. Chakma, Performance analysis of a compact dual-mode antenna operating at UWB and ISM band for wireless medical applications, in *2018 International Conference on Innovations in Science, Engineering and Technology (ICISET), Chittagong, Bangladesh*, vol. 2018, pp. 211–216. https://doi.org/10.1109/ICISET.2018.8745633
19. Online link: https://www.avenga.com/magazine/all-you-need-to-know-about-iomt/
20. X. Tong, C. Liu, X. Liu, H. Guo, X. Yang, Switchable ON-/OFF-body antenna for 2.45 GHz WBAN applications. IEEE Trans. Antennas Propag. **66**(2), 967–971 (2018). https://doi.org/10.1109/TAP.2017.2780984
21. Q.H. Dang, N. Nguyen-Trong, S.J. Chen, C. Fumeaux, Frequency-reconfigurable UHF wearable textile antenna for RFID applications, in *2023 5th Australian Microwave Symposium (AMS), Melbourne, Australia*, vol. 2023, pp. 23–24. https://doi.org/10.1109/AMS57822.2023.10062314
22. S.M. Abbas, Y. Ranga, K.P. Esselle, Reconfigurable antenna options for 2.45/5 GHz wireless body area networks in healthcare applications, in *2015 37th Annual International Conference*

of the IEEE Engineering in Medicine and Biology Society (EMBC), Milan, Italy, vol. 2015, pp. 5465–5468. https://doi.org/10.1109/EMBC.2015.7319628

23. X.-T. Yang, H. Wong, J. Xiang, Polarization reconfigurable planar inverted-F antenna for implantable telemetry applications. IEEE Access **7**, 141900–141909 (2019). https://doi.org/10.1109/ACCESS.2019.2941388
24. U. Musa et al., Recent advancement of wearable reconfigurable antenna technologies: A review. IEEE Access **10**, 121831–121863 (2022). https://doi.org/10.1109/ACCESS.2022.3222782
25. Z.X. Oh, K.H. Yeap, C.S. Voon, K.C. Lai, P.C. The, A multiband antenna for biomedical telemetry and treatments. J. Phys. Conf. Ser. **1502**, 012013 (2020). https://doi.org/10.1088/1742-6596/1502/1/012013
26. N. Kaur, J.S. Sivia, Rajni, Design of frequency reconfigurable planar antenna using artificial neural network. Int. J. Microw. Wirel. Technol. **14**, 1107–1118 (2022)
27. M. Al-gburi, M. Ilyas, A novel design reconfigurable antenna based on the metamaterial for wearable applications. J. Phys. Conf. Ser. **1973**, 012042 (2021)
28. S. Ahmad, B. Manzoor, S. Naseer, N. Santos-Valdivia, A. Ghaffar, M.I. Abbasi, X-shaped slotted patch biomedical implantable antenna for wireless communication networks. Wirel. Commun. Mob. Comput. **2022**, 7594587 (2022)
29. S.-E. Didi, I. Halkhams, A. Es-saqy, M. Fattah, S. Mazer, M. El Bekkali, Creation of a soft circular patch antenna for bio-medical applications for 5G at frequency 2.45 GHz. Results Eng. **22**, 102319 (2024)
30. M. Salim, A. Pourziad, A novel reconfigurable spiral-shaped monopole antenna for biomedical applications. Prog. Electromagn. Res. Lett. **57**, 79–84 (2015)

Chapter 10
Future Trends and Emerging Antenna Technologies in Biomedical Applications

Contents

© The Author(s), under exclusive license to Springer Nature Switzerland AG 2026

A. Segun Ayokunle et al., *Biomedical Antenna Technology*,
https://doi.org/10.1007/978-3-031-95268-5_10

10.1 Current State of Biomedical Antenna Technology

Biomedical antenna technology has experienced significant advancements, driven by the increasing need for continuous health monitoring, telemedicine, personalized healthcare, and minimally invasive medical treatments. Biomedical systems, particularly those relying on wireless communication for data transfer, have transformed the way medical practitioners and patients interact. The role of antennas in this technological evolution is foundational, enabling the transfer of data, energy, and signals across various biomedical applications, from wearable devices to medical implants and diagnostic equipment. This chapter investigates into the state-of-the-art developments in biomedical antenna technology, with a focus on key advancements, ongoing challenges, and emerging opportunities within the field [1].

As healthcare becomes more personalized and patient-centric, medical devices are evolving to incorporate real-time data analytics and remote monitoring capabilities. A surge in research aimed at improving the functionality of biomedical antennas in body-centric environments is reshaping the healthcare landscape. The human body presents unique challenges for antenna design, influencing how antennas are integrated into health devices that must perform efficiently while remaining comfortable and non-intrusive. The following subsections will explore the various applications of biomedical antennas, such as wearable health devices, medical implants, medical imaging systems, and therapeutic treatments, as well as the current state of these technologies.

10.1.1 Wearable Health Monitoring Devices

The global rise of wearable health devices represents a major innovation in modern healthcare. These devices include consumer-oriented wearables like smartwatches and fitness trackers, as well as medical-grade devices such as continuous glucose monitors (CGMs), electrocardiogram (ECG) sensors, and blood pressure cuffs. These devices are integral to real-time health monitoring, offering individuals valuable insights into their health and enabling clinicians to remotely track and analyze patient data. As a result, wearable health devices are contributing to the paradigm shift toward more personalized and preventative healthcare approaches [2].

At the core of these devices is the antenna, responsible for wirelessly transmitting data between the device and external systems, such as smartphones, healthcare cloud platforms, or hospital monitoring equipment. Over the years, microstrip patch antennas have gained prominence in wearable health devices due to their efficient design, thin profile, and lightweight properties. These antennas are highly suitable for integration into small and slim form factors, ensuring that wearables remain comfortable for long-term use without sacrificing antenna performance.

Microstrip patch antennas excel at supporting short-range communication protocols, such as Bluetooth, Wi-Fi, Zigbee, and low-energy Bluetooth, which are commonly used in wearable health devices. However, the increasing complexity of wearable devices has led to a growing demand for antennas that support high-speed data transmission and long-range communication for telemedicine and remote healthcare applications. The rollout of 5G networks, for instance, is expected to further enhance the capabilities of wearable devices, allowing for seamless communication between patients and healthcare providers even in remote areas.

As wearable health devices are designed to operate in close proximity to the human body, antenna performance can be affected by the lossy nature of biological tissue, which can cause signal attenuation and interference. Researchers are dedicated to overcoming this issue by developing new materials, such as metamaterials, and optimizing antenna placement and design. These advanced materials have unique electromagnetic properties that can help mitigate the effects of body tissue, improving antenna efficiency and signal strength [3].

Another key challenge in wearable health devices is miniaturization. As these devices become more sophisticated, they must include advanced sensors and multiple communication protocols without increasing in size. Achieving miniaturization without compromising on the efficiency and functionality of antennas requires a careful balance between size, power consumption, and signal performance. To achieve this, researchers are exploring novel antenna architectures, including reconfigurable antennas that can dynamically adjust their properties based on environmental conditions.

10.1.2 Medical Implants

Medical implants are another area where biomedical antenna technology plays a critical role. These implants are embedded within the body for therapeutic purposes, such as pacemakers, neural probes, cochlear implants, and drug delivery systems see Fig. 10.1. In many cases, medical implants rely on antennas to wirelessly communicate with external devices for remote monitoring, diagnostics, and programming. For example, pacemakers can transmit data to a remote monitoring system, allowing healthcare providers to track the patient's heart rate and adjust device settings without requiring in-person visits.

Implantable antennas must contend with several limitations, including size, biocompatibility, and signal quality. Because the space within the human body is limited, antenna designs must be highly compact without sacrificing their ability to transmit and receive signals efficiently. In addition to being miniaturized, implantable antennas must be made of biocompatible materials that will not induce adverse reactions in the body. Common biocompatible materials include titanium, ceramics, and special polymers, all of which can withstand the harsh conditions of the body and prevent rejection or inflammation [4].

Moreover, the human body itself presents challenges to signal transmission. Tissue absorption and interference from surrounding biological structures can degrade the performance of implantable antennas. To address these issues,

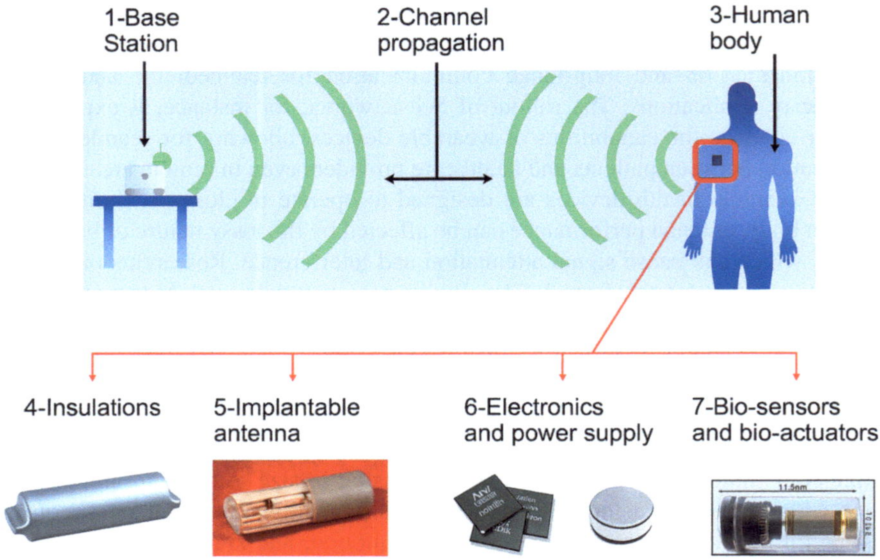

Fig. 10.1 Medical implant system [4]

researchers are focused on optimizing the antenna design to minimize signal attenuation and maximize efficiency. Advanced materials and coating techniques are being explored to improve the performance of implantable antennas while ensuring their safety and durability.

Another important consideration for implantable antennas is power efficiency. Many medical implants rely on batteries, and reducing energy consumption can extend the lifetime of these devices. Power efficiency can be improved by optimizing antenna designs and implementing energy harvesting technologies, such as piezoelectric systems or thermoelectric generators, which can capture and convert energy from the surrounding environment.

10.1.3 Medical Imaging Systems

Medical imaging systems, such as magnetic resonance imaging (MRI), ultrasound, and computed tomography (CT), have revolutionized the diagnosis and treatment of various health conditions. These systems rely on antennas to generate and detect electromagnetic signals or sound waves, which are used to create detailed images of internal body structures. Antennas play a vital role in the functioning of these imaging systems, influencing the quality of the images produced and the accuracy of diagnoses [5].

In MRI, antennas are responsible for transmitting and receiving radiofrequency (RF) signals, which are used to excite hydrogen nuclei within the body's tissues.

These signals are then captured and processed to create high-resolution images of internal organs and tissues. MRI antennas must be designed to operate within specific frequency ranges to ensure optimal performance while minimizing interference from external sources.

Ultrasound imaging systems, which use high-frequency sound waves to generate real-time images of tissues and organs, also rely on antennas to generate and receive these waves. The design of ultrasound antennas is critical to ensuring accurate imaging, as the frequency and power of the sound waves must be precisely controlled. Furthermore, the ability of ultrasound antennas to transmit signals at the appropriate angles and distances is essential for obtaining clear images of internal structures.

In both MRI and ultrasound systems, the design of antennas must take into account factors such as signal strength, frequency response, and radiation patterns. These antennas must also be robust enough to withstand the conditions of the medical environment, where exposure to heat, moisture, and electrical noise is common. The ongoing development of advanced materials, such as superconductors and flexible substrates, is helping to improve the performance and reliability of medical imaging antennas.

10.1.4 Therapeutic Applications

In therapeutic applications, antennas are used to deliver energy to specific areas of the body for treatments such as microwave ablation, radiofrequency ablation, and transcutaneous electrical nerve stimulation (TENS). These treatments are used to target tissues and organs for various purposes, including the destruction of cancerous cells, pain relief, and tissue regeneration. Antennas are used to focus and direct energy precisely to the treatment site, ensuring that the surrounding healthy tissue is not harmed [6].

In microwave ablation therapy, antennas generate microwave energy that is directed at a tumor, causing it to heat up and destroy cancerous cells. These antennas must be carefully designed to ensure that the energy is delivered to the tumor in a controlled and focused manner. Similarly, in radiofrequency ablation, antennas are used to deliver energy to tissues for the purpose of removing or shrinking tumors or treating abnormal heart rhythms.

The antenna design for therapeutic applications must consider factors such as energy delivery efficiency, power consumption, and safety. Researchers are working on developing antennas with better power control, more precise energy delivery, and lower heat dissipation to improve treatment outcomes and reduce the risk of adverse effects.

10.2 Promising Advances and Research Directions

As biomedical applications continue to evolve and expand, antenna technology plays a pivotal role in facilitating the functionality of these innovations. From wearable health devices to advanced medical implants and diagnostic systems, antennas are at the core of enabling effective communication, energy transfer, and real-time monitoring. Several promising advances are being explored in the field of biomedical antenna technology, which will allow healthcare solutions to become more reliable, user-friendly, and precise. These advancements hold great promise for enhancing the quality of patient care, improving health outcomes, and paving the way for the next generation of biomedical devices [7].

In this section, we delve into some of the most promising developments in biomedical antenna technology, covering biocompatible and flexible materials, miniaturization, wearable body area networks (BANs), microwave and terahertz technologies, and wireless power transfer (WPT) systems for implants. Each of these innovations represents a major breakthrough in overcoming the limitations faced by traditional antenna technologies and opens new possibilities for their application in healthcare.

10.2.1 Biocompatible and Flexible Materials

A major challenge in the design of biomedical antennas is ensuring that they are biocompatible, meaning they must not cause any harmful reactions in the body, while also being effective at transmitting and receiving signals. For wearable devices, which come into direct contact with the skin, and for implantable devices, which must operate within the body, the choice of materials is crucial for both safety and performance. Recent developments in flexible and biocompatible materials have opened up new opportunities in this area.

10.2.1.1 Flexible Substrates for Wearable and Implantable Devices

Materials such **as** polydimethylsiloxane (PDMS)**,** polyethylene terephthalate (PET), and organic polymers have been investigated for their potential to serve as flexible substrates for biomedical antennas. PDMS, a silicon-based material, is widely used due to its excellent biocompatibility, flexibility, and ease of fabrication. These flexible substrates can conform to the body's curves, enabling seamless integration into wearable devices such as health monitors, fitness trackers, and wearable ECG sensors. Their ability to stretch and bend without compromising antenna performance makes them ideal for applications where the device will be subject to continuous motion or body deformation [8].

10.2.2 Conductive Materials for Enhanced Performance

The advancement of conductive materials, particularly graphene and conductive polymers, has significantly impacted antenna performance in flexible and stretchable biomedical applications. Graphene, a single layer of carbon atoms arranged in a two-dimensional lattice, possesses remarkable electrical conductivity and mechanical strength while remaining extremely thin and lightweight. When used in antennas, graphene can maintain signal integrity, even when the antenna undergoes deformation due to body movement. Conductive polymers, such as poly(3,4-ethylenedioxythiophene) (PEDOT), offer similar advantages in terms of flexibility, and they can be fabricated through low-cost, scalable processes, making them attractive for mass production of biomedical devices.

Research is ongoing to explore how these materials can be optimized to balance the trade-off between flexibility, durability, and electrical performance. For instance, there is a growing interest in developing composite materials that combine flexible polymers with metal nanoparticles or graphene to enhance conductivity while maintaining the mechanical properties needed for wearable and implantable applications [9].

10.2.3 Miniaturization and Integration

One of the most significant trends in biomedical antenna design is miniaturization, particularly for implantable medical devices where space is constrained. The shrinking of antenna size without sacrificing performance is crucial for ensuring that devices can be both small and comfortable for patients. Advances in nano-structured antennas and the use of metamaterials and fractal geometries hold the potential to drastically reduce antenna dimensions while maintaining or even improving performance.

10.2.3.1 Nano-Structured and Metamaterial Antennas

Metamaterials, engineered materials with properties not found in nature, offer significant advantages in antenna design. They can manipulate electromagnetic waves in ways that traditional materials cannot, allowing for the creation of compact antennas that can operate efficiently at small scales. Researchers are experimenting with nano-structured antennas that utilize metamaterials to reduce size and improve performance, especially for applications requiring high-frequency signals, such as medical imaging and wireless data transfer [10].

10.2.3.2 Fractal Antennas

Fractal antennas, characterized by self-repeating structures at different scales, are another promising approach to miniaturization. These antennas can operate at multiple frequencies and offer high performance in small packages, making them particularly suitable for medical implants that require both small size and broadband capabilities. The use of fractal geometries enables antennas to be highly resonant, reducing the need for large physical dimensions [11].

10.2.3.3 Integration with Smart Textiles

Beyond miniaturization, integrating antennas directly into smart textiles represents an exciting new direction. Smart textiles are fabrics that incorporate sensors and communication systems, making them capable of monitoring physiological parameters such as heart rate, body temperature, and blood oxygen levels. These textiles can be worn like everyday clothing, offering a non-intrusive and comfortable solution for continuous health monitoring. Antennas embedded into these fabrics need to be flexible, lightweight, and capable of transmitting data wirelessly to external devices such as smartphones or medical hubs. Research is progressing to develop seamless integration techniques that do not compromise the performance or comfort of the wearer [12].

10.2.4 Wearable Body Area Networks (BANs)

The Body Area Network (BAN) concept refers to a system of interconnected devices worn on or near the body that communicates wirelessly to monitor various health parameters in real time. The evolution of **wireless** communication technologies including Bluetooth Low Energy (BLE), Zigbee, and Wi-Fi has played a pivotal role in the development of wearable devices. Antennas used in these systems must support short-range communications, while maintaining low-power consumption to ensure the longevity of the device [13].

10.2.4.1 Low-Power Consumption and Short-Range Communication

One of the primary considerations in BAN antenna design is optimizing **for** low-power consumption, which is critical for the practical use of wearable health devices. Since wearable devices are typically battery-powered, it is important that the antennas are able to transmit and receive signals efficiently while consuming minimal energy. Research is focused on designing ultra-low power antennas that operate in the Industrial, Scientific, and Medical (ISM) frequency bands, such as

2.4 GHz and 5.8 GHz, which are commonly used for wireless communication in medical devices [14].

10.2.4.2 Optimizing Performance near the Human Body

A significant challenge in designing antennas for wearable devices is the interaction between the antenna and the human body. The human body is conductive and can significantly alter the propagation of electromagnetic waves. Antennas designed for body-centric communication must be able to operate efficiently despite the varying impedance caused by the body, which can distort signal quality and reduce transmission range. Researchers are investigating body-coupled antennas and antenna designs that can compensate for this distortion, allowing for more reliable communication in body-centric environments [15].

10.2.4.3 Ultra-Wideband (UWB) Antennas

In addition to traditional short-range frequencies, there is increasing interest in ultra-wideband (UWB) communication for medical applications. UWB technology, which operates over a broad frequency range, offers high data transfer rates and the ability to penetrate various materials, including human tissues. This technology could be particularly useful for non-invasive monitoring of vital signs, such as respiratory rate, blood pressure, and blood glucose levels, as well as for tracking medical devices within the body.

10.2.5 Microwave and Terahertz Technologies for Non-invasive Monitoring

The use of high-frequency electromagnetic waves, particularly in the microwave and terahertz (**THz**) ranges, has gained attention in the medical field due to its non-invasive nature and potential for deep tissue penetration without causing harm to the body. These technologies have shown promise in applications such as medical imaging, cancer detection, and tissue characterization. Terahertz radiation, in particular, has the ability to penetrate biological tissues and provide detailed information about their structure and composition, making it an invaluable tool for diagnostics and treatment [16].

10.2.5.1 Microwave and THz Imaging

Microwave and terahertz imaging are emerging as powerful tools for diagnosing various diseases, including cancer, diabetes, and neurodegenerative disorders. By using antennas designed to transmit and receive these high-frequency electromagnetic waves, researchers can obtain real-time images of the inside of the body without the need for invasive procedures. For example, THz imaging has been explored for detecting skin cancer and breast cancer, with promising results due to the sensitivity of THz waves to tissue abnormalities.

10.2.5.2 Challenges in THz Antenna Design

Designing antennas that can efficiently transmit and receive terahertz waves presents several challenges, including low-power efficiency, high conversion losses, and the need for highly directional radiation patterns. Additionally, the small wavelength of THz radiation requires precise antenna design and alignment to achieve the necessary resolution for medical applications. Ongoing research is focused on developing compact THz antennas that can be integrated into portable diagnostic systems, which could significantly improve the accessibility of THz-based imaging technologies.

10.2.6 Wireless Power Transfer (WPT) for Implants

As medical implants become more advanced, the need for efficient and continuous power supply systems has become increasingly important. Many implants, such as cochlear implants, neurostimulators, and drug delivery systems, require a constant power source to function optimally. Traditional battery-based systems have limitations in terms of size, longevity, and the need for periodic replacements or recharging. Wireless power transfer (WPT) offers a promising alternative, allowing for efficient power delivery to implants without the need for invasive surgeries or battery replacements [17].

10.2.6.1 Resonant Inductive Coupling for WPT

One of the most widely studied methods of wireless power transfer is resonant inductive coupling, where energy is transferred through magnetic fields between two coils tuned to resonate at the same frequency. This technology has been successfully used in applications such as **wireless** charging for medical devices, including implants. Researchers are working to improve the efficiency and range of these inductive power transfer systems, making them more suitable for use in implantable biomedical devices.

10.2.6.2 Hybrid WPT System

Another exciting area of research is the development of hybrid wireless power transfer systems, which combine multiple energy transfer methods, such as inductive coupling, **capacitive coupling**, and radiative transfer. These hybrid systems aim to overcome the limitations of traditional WPT systems, such as poor efficiency or short range, and provide more reliable power delivery to implants [18].

10.2.6.3 Energy Harvesting for Implants

In addition to wireless power transfer, energy harvesting techniques are being explored as a way to power biomedical implants. These methods involve capturing energy from the surrounding environment such as body heat, motion, or vibration and converting it into usable electrical energy. While energy harvesting systems have the potential to reduce the need for external power sources, their efficiency is still an area of active research. The field of biomedical antenna technology is experiencing rapid advancements, driven by the increasing demand for wearable health devices, medical implants, and diagnostic systems. Innovations in biocompatible and flexible materials, miniaturization, wearable body area networks, microwave and terahertz technologies, and **wireless** power transfer systems are transforming healthcare delivery, offering new possibilities for non-invasive monitoring, real-time data transmission, and improved patient outcomes.

As research continues to address the challenges of miniaturization, biocompatibility, and efficiency, the potential for biomedical antennas to enhance healthcare systems and contribute to the development of more personalized, accessible, and effective treatments will continue to grow. These advancements are setting the stage for the next generation of medical devices that will play a crucial role in shaping the future of healthcare [19].

10.3 Envisioning the Next Generation of Healthcare Antennas

The evolution of healthcare technology has already begun to redefine the landscape of medical care. As we move into the future, healthcare antennas will be at the forefront of innovations that combine cutting-edge technology with everyday life. The next generation of healthcare antennas will not only enhance the performance of healthcare devices but will also make them smaller, more efficient, and seamlessly integrated into daily life, thereby contributing to an era of personalized, real-time healthcare monitoring.

In this section, we explore the future trends in healthcare antennas and how they will shape the field of medical devices, patient care, and diagnostics. The future will likely see the intersection of multiple disciplines, including antenna design, materials science, nanotechnology, and biotechnology. Below, we outline several key trends that are expected to guide the development of healthcare antennas in the coming decades [20].

10.3.1 Seamless Integration with the Human Body

As healthcare technology continues to advance, one of the most exciting prospects is the seamless integration of antennas with the human body. Traditional healthcare devices are often bulky and cumbersome, requiring a significant physical footprint and limiting user comfort. Future healthcare antennas will be embedded into wearable and implantable devices that are flexible, lightweight, and unobtrusive, ensuring that they do not interfere with the user's daily activities or comfort.

10.3.1.1 Wearable and Implantable Devices

A growing trend in the healthcare industry is the shift toward wearable health devices, such as smartwatches, fitness trackers, and health monitoring patches. These devices are expected to incorporate highly efficient antennas that enable continuous health monitoring without being noticeable or uncomfortable. The next generation of antennas will be integrated into these wearables with flexibility and biocompatibility in mind. For instance, antenna designs may be optimized to conform to the skin or even directly to tissues, ensuring maximum comfort and minimal disruption.

Implantable healthcare devices, such as pacemakers, insulin pumps, and neural stimulators, will also see a shift toward the use of integrated antennas. These devices will be capable of transmitting data to external monitoring systems, allowing healthcare providers to track the patient's condition in real time. The integration of these antennas within the body will facilitate long-term, continuous monitoring without the need for external equipment or the inconvenience of regular checkups [21].

10.3.1.2 Innovations in Materials Science and Biocompatibility

To support the development of these integrated antennas, innovations in materials science will be key. Materials that are biocompatible, lightweight, and flexible will be crucial in designing antennas that can be embedded directly into the body or skin. Researchers are exploring materials such as conductive polymers, flexible substrates, and biodegradable electronics that are safe for long-term use in the body.

Antennas that are placed on the skin will also need to adhere to the body's natural curves and movements without causing irritation. Innovations in stretchable and self-healing materials will be critical in ensuring that these devices remain functional over extended periods of time, while still maintaining their high performance characteristics [22].

10.3.2 Multifunctional Antennas

As wearable health devices become increasingly sophisticated, the need for multifunctional antennas will grow. The next generation of healthcare antennas will not only focus on improving performance in one area but will be designed to serve multiple purposes simultaneously.

10.3.2.1 Data Transmission, Location Tracking, and Wireless Power

For example, an advanced healthcare antenna could simultaneously transmit health data, track the user's location, and provide wireless power to the device. These capabilities could drastically reduce the need for multiple separate antennas and minimize the size and complexity of the overall device. In addition, a multifunctional antenna can improve efficiency by enabling these various functionalities to work in harmony without interfering with one another.

Healthcare devices that integrate GPS tracking, health data monitoring, and wireless charging can be combined into a single compact system, which will not only be more efficient but also more convenient for users. The ability to track patients' locations—whether in a hospital setting or through remote monitoring at home—could play a critical role in managing patients' conditions, ensuring that they receive the appropriate level of care [22].

10.3.2.2 Flexible and Adaptive Designs

Multifunctional antennas will also need to be adaptable to different environments and requirements. For instance, healthcare devices worn by individuals with chronic conditions may need to continuously monitor vital signs, while those used for emergency response may need to transmit data over long distances and function in various environmental conditions.

Designers will likely turn to adaptive and reconfigurable antenna technologies that can dynamically alter their operating parameters based on specific use cases. This would allow a single device to optimize its functionality for different purposes, reducing the need for separate devices and antennas for each task [23].

10.3.3 Smart Textiles and Implantable Wearables

In recent years, there has been a surge in research focused on smart textiles, which integrate antennas and sensors directly into fabrics. These textiles could revolutionize healthcare by enabling continuous, non-invasive health monitoring in a manner that is virtually invisible to the user. Smart textiles will play a significant role in embedding healthcare antennas within everyday clothing, making them discreet and comfortable to wear.

10.3.3.1 Integration with Fabrics

Smart fabrics will not only be used for monitoring vital signs such as heart rate, temperature, and blood oxygen levels, but they could also enable real-time diagnostics for conditions like diabetes or sleep apnea. These fabrics will be embedded with miniature sensors and antennas, allowing for the seamless collection and transmission of health data to external devices or healthcare providers.

Such textiles could be incorporated into clothing items that are worn throughout the day, including shirts, pants, and undergarments. These fabrics will be designed to be durable, breathable, and washable, providing long-term functionality without sacrificing comfort or practicality [24].

10.3.3.2 Implantable Wearables

Beyond textiles, implantable wearables such as smart contact lenses or ingestible sensors will also benefit from advancements in antenna technology. These devices will be capable of monitoring a range of health parameters from within the body, providing real-time data directly from the source. The miniaturization of antennas will be essential in making these devices feasible and effective.

For instance, smart contact lenses could monitor eye health, detect early signs of disease, or even measure blood sugar levels for diabetic patients. Ingestible sensors could track gastrointestinal health or monitor drug delivery systems within the body. These advancements would allow for unprecedented levels of personalized healthcare, where data is continuously gathered and analyzed to provide the most accurate and timely information about a person's health [25].

10.3.4 Enhanced Imaging and Diagnosis

One of the most promising areas for healthcare antennas is the enhancement of medical imaging and diagnostic techniques. Antennas will play a critical role in advancing technologies such as terahertz (THz) and microwave imaging, enabling higher-resolution, non-invasive imaging methods that can be used to detect diseases earlier with greater accuracy.

10.3.4.1 Terahertz and Microwave Imaging

THz and microwave imaging technologies are poised to revolutionize medical diagnostics by offering enhanced resolution and contrast compared to traditional imaging techniques like X-rays or MRIs. These technologies will rely heavily on advanced antenna designs that can operate at these high-frequency ranges while maintaining efficiency and minimizing interference.

For instance, terahertz imaging systems have shown promise in identifying early-stage cancers, including skin, breast, and lung cancers, by detecting changes in tissue composition before they become visible through conventional imaging techniques. Similarly, microwave imaging could be used to detect conditions like brain tumors or neurological disorders, where early detection is critical for effective treatment [26].

Antennas designed for these imaging systems will need to be highly efficient, compact, and capable of handling the high-frequency signals required for accurate imaging. Researchers are also exploring the use of metamaterials and novel antenna structures to improve the performance of these systems, pushing the boundaries of what is possible in non-invasive medical diagnostics.

10.3.4.2 Real-Time Diagnostic Tools

The miniaturization of antennas will allow for real-time imaging and diagnostic tools that can be used at the point of care, whether in a doctor's office, a hospital, or even at home. These portable, easy-to-use devices will enable healthcare professionals to quickly assess a patient's condition and make informed decisions without the need for time-consuming or costly lab tests [27].

This level of accessibility will not only improve patient outcomes by enabling faster diagnoses but will also reduce healthcare costs by streamlining the diagnostic process. Additionally, these portable devices could be used in underserved or remote areas, where access to advanced diagnostic equipment may be limited, ensuring that everyone has access to high-quality healthcare [28].

The next generation of healthcare antennas will transform the way we think about medical devices, patient care, and diagnostics. From seamless integration with the human body to multifunctional antennas, smart textiles, implantable wearables, and enhanced imaging technologies, these innovations will significantly improve the quality of healthcare while making it more accessible and efficient. As antenna designs continue to evolve, the possibilities for healthcare applications are virtually limitless, paving the way for a future where real-time, personalized healthcare is available to all [29].

10.4 Open Research Challenges in Biomedical Antenna Technology

Biomedical antenna technology is a rapidly growing field with the potential to revolutionize healthcare systems. Antennas are pivotal in enabling communication between implanted devices and external systems, allowing for the transmission of vital health information. However, despite significant advances in this field, there remain several challenges that need to be addressed for the technology to reach its full potential. These challenges are primarily related to power efficiency, antenna performance near the human body, biocompatibility and safety, and cost-effective manufacturing. Each of these challenges presents unique opportunities for research, and solving them will be crucial for the continued advancement and implementation of biomedical antenna technologies.

10.4.1 Power Efficiency

Power efficiency is one of the most pressing challenges in the design of biomedical antennas, especially those intended for implantation within the human body. Implantable devices, such as pacemakers, neurostimulators, and drug delivery systems, require a constant source of power to maintain their functionality. The performance of these devices is heavily dependent on the energy provided by their batteries. However, there is an inherent trade-off between power consumption and device size. Batteries must be small enough to be implanted without causing discomfort or requiring major surgical intervention, but they must also be powerful enough to support the device's functions over an extended period [30].

As battery size and capacity are limited, the challenge lies in maximizing the power efficiency of these devices while minimizing the energy consumption of the antennas used for communication. Biomedical antennas often operate in highly constrained environments, where the available power may be further limited by the patient's health condition or other medical devices in use. Therefore, enhancing power efficiency is crucial for the longevity and reliability of implantable devices.

Furthermore, the ability to harvest energy from the surrounding environment or wirelessly transfer power to the implantable devices would significantly reduce the dependency on traditional batteries. Energy harvesting methods such as piezoelectric, thermoelectric, or solar power could be integrated with antennas to provide a more sustainable power source. Additionally, wireless power transfer (WPT) techniques could allow energy to be transmitted to the implanted device from an external source, eliminating the need for frequent battery replacements. These advancements would not only extend the operational lifetime of the devices but also improve patient safety by reducing the risk of complications associated with battery failure or replacement surgeries. Research into these energy-efficient solutions is essential to address the challenges posed by limited power availability in implantable devices.

10.4.2 Antenna Performance Near the Human Body

Another significant challenge in biomedical antenna design is ensuring that the antenna's performance is not compromised when operating in proximity to the human body. The human body, composed of various tissues such as skin, muscle, fat, and bone, has a complex and variable electromagnetic response. When antennas are placed near or within the human body, the performance of the antenna can be severely affected due to issues such as signal attenuation, detuning, and energy absorption. These phenomena can lead to reduced communication range, lower data transmission rates, and overall compromised functionality [31].

A major concern is signal attenuation, which occurs when electromagnetic signals are absorbed by the body's tissues, leading to a reduction in the strength of the transmitted signal. This attenuation is more pronounced at certain frequencies, particularly in the radio frequency (RF) spectrum, which is commonly used for wireless communication in biomedical applications. Moreover, the human body can cause detuning of the antenna, which occurs when the impedance matching between the antenna and the surrounding medium is altered due to body-induced changes. Detuning can result in reduced antenna efficiency and signal degradation [32].

To address these issues, researchers are exploring the development of body-centric communication models that take into account the unique properties of the human body. These models could help in designing antennas that can adapt to the varying conditions created by the body. Adaptive antennas, which can automatically adjust their characteristics in response to changes in their environment, are one promising solution. Such antennas could adjust their radiation pattern, frequency, or impedance to compensate for the variations induced by the human body, thus maintaining stable performance even when the antenna is embedded in or near the body.

Moreover, the integration of antennas with flexible and stretchable materials is another avenue of research that could enhance antenna performance in biomedical applications. Flexible antennas are capable of conforming to the body's shape, ensuring minimal disruption to signal transmission. Additionally, these antennas can be embedded directly into medical devices or implanted into the body, offering a seamless integration that maintains functionality without compromising patient comfort. Research into optimizing antenna designs for body-centric communication will be key to improving the performance of biomedical antennas in real-world scenarios [33].

10.4.3 Biocompatibility and Safety

Ensuring that biomedical antennas are biocompatible is critical, as they are often implanted in or come into direct contact with human tissues for extended periods. The materials used to construct these antennas must not cause any adverse biological reactions or toxicity when exposed to bodily fluids or tissues. This is especially important for long-term implants, as they can remain inside the body for years without requiring removal or replacement.

Biocompatibility involves ensuring that the materials used for the antenna are non-toxic, non-carcinogenic, and do not provoke an immune response. Antennas must also be resistant to corrosion or degradation caused by bodily fluids, ensuring that the device continues to function properly over time. Additionally, the surface of the antenna must be smooth and non-irritating to prevent inflammation or other complications.

Future research will need to focus on the development of novel materials that combine both electrical efficiency and biocompatibility. Materials such as biocompatible metals, polymers, and ceramics are being investigated for their potential to meet these requirements. For example, titanium and its alloys are commonly used for implantable medical devices due to their excellent biocompatibility, strength, and resistance to corrosion. Similarly, polymers such as polyimide and polytetrafluoroethylene (PTFE) are being explored for their flexibility, durability, and biocompatibility, making them suitable for use in flexible or wearable biomedical antennas [34].

In addition to material development, the design of antennas that can minimize tissue irritation and inflammation is another area of focus. The antenna's shape, size, and method of implantation must be carefully considered to avoid excessive pressure on surrounding tissues or blood vessels. Furthermore, research into non-invasive techniques for antenna implantation, such as through minimally invasive surgery or injection-based methods, could significantly reduce the risks associated with traditional surgical procedures.

10.4.4 Cost and Manufacturing

Despite the tremendous potential of biomedical antenna technology, the cost of manufacturing these devices remains a significant barrier to widespread adoption. The production of advanced biomedical antennas, particularly those designed for small, implantable devices, is often expensive due to the complexity of the designs, the materials used, and the specialized fabrication techniques required. This high cost is a major challenge for the scalability of these technologies, as it limits their accessibility to a broader range of patients, particularly in low-resource settings.

To make biomedical antennas more affordable and accessible, research into cost-effective manufacturing methods is essential. One promising approach is the use of

3D printing, which offers the ability to create highly intricate designs with reduced material waste and lower production costs. 3D printing also allows for the customization of antennas to meet specific patient needs, such as tailoring the antenna's size and shape to fit individual anatomical requirements. Additionally, roll-to-roll processing, which is commonly used in the production of flexible electronics, could also be leveraged to produce biomedical antennas at a lower cost and in larger quantities [35].

The development of these cost-effective manufacturing techniques would help drive the widespread adoption of biomedical antenna technologies, making them more accessible to healthcare providers and patients alike. Furthermore, advancements in manufacturing processes could also enable the production of more complex and functional antennas, which would enhance the overall performance and capabilities of biomedical devices.

In addition to manufacturing costs, there is a need for standardized testing and certification procedures to ensure the quality and safety of biomedical antennas. The establishment of industry-wide standards for performance, biocompatibility, and reliability would help streamline the approval process for new devices, reducing the time and cost required to bring these technologies to market [36].

Biomedical antenna technology has made significant strides in recent years, offering promising solutions for the communication needs of implantable medical devices. However, several key challenges remain in the areas of power efficiency, antenna performance near the human body, biocompatibility and safety, and cost-effective manufacturing. Addressing these challenges will require continued research and innovation across multiple disciplines, including materials science, electrical engineering, biomedical engineering, and manufacturing technologies.

Advancements in energy harvesting, wireless power transfer, body-centric communication models, and adaptive antennas hold the potential to overcome many of the current limitations of biomedical antenna systems. Additionally, the development of biocompatible materials and cost-effective manufacturing techniques will help make these technologies more accessible and reliable for widespread clinical use. As research in these areas progresses, it is expected that biomedical antenna technology will play an increasingly vital role in improving healthcare outcomes, particularly in the management of chronic conditions, monitoring of vital signs, and the delivery of personalized medical treatments.

10.5 Summary: The Future of Biomedical Antenna Technology

Biomedical antenna technology is poised to play a pivotal role in the future of healthcare. Driven by remarkable advances in materials science, miniaturization, and communication technologies, the development of biomedical antennas has the potential to revolutionize the way we monitor, diagnose, and treat a wide range of medical conditions. The increasing demand for wearable and implantable health monitoring systems has made antennas essential components for enabling wireless communication and data transmission between medical devices and external systems. Whether used in wearable fitness trackers, implantable pacemakers, or advanced diagnostic tools, antennas are key to enhancing the functionality and efficacy of medical devices. However, as promising as the future of biomedical antenna technology may be, several challenges remain that need to be addressed before this technology can achieve its full potential.

10.5.1 The Role of Biomedical Antennas in Future Healthcare

The need for advanced healthcare systems is rapidly growing, driven by an aging population, an increasing prevalence of chronic diseases, and the widespread adoption of personalized medicine. To address these challenges, the healthcare industry is increasingly relying on remote monitoring and telemedicine solutions that can track patients' health data in real time. Biomedical antennas are at the heart of these innovations, enabling the transmission of critical health data from wearable devices, implantable medical devices, and diagnostic equipment to healthcare professionals, caregivers, or even the patients themselves.

Antennas are the primary components that allow these devices to communicate wirelessly, enabling them to transmit data such as heart rate, blood pressure, glucose levels, and oxygen saturation levels. Wireless communication eliminates the need for cumbersome wired connections, offering patients more freedom, comfort, and mobility. For instance, in the case of implantable medical devices like pacemakers or neurostimulators, antennas are integral in facilitating communication with external devices for real-time monitoring and adjustment of device settings. Moreover, the use of biomedical antennas in wireless health monitoring systems has the potential to greatly improve healthcare outcomes by enabling timely interventions, reducing hospital visits, and improving patient quality of life.

Miniaturization has allowed for the development of smaller, more efficient antennas that can be easily integrated into medical devices without compromising the device's size or functionality. The continuous progress in material science has further enabled the design of antennas that are not only smaller but also more flexible, durable, and capable of operating effectively in a variety of environments. As a result, we are seeing a new era of wearable and implantable devices that are more comfortable, more powerful, and more reliable than ever before.

10.5.2 Challenges to Overcome in Biomedical Antenna Development

Despite these exciting advancements, there are still several key challenges that need to be addressed before biomedical antenna technology can reach its full potential. Among these challenges, the most pressing include power efficiency, biocompatibility, integration with the human body, and cost-effective manufacturing. These obstacles are interrelated and require innovative solutions from various fields of research.

10.5.2.1 Power Efficiency

Power efficiency remains one of the most critical challenges in the development of biomedical antennas, particularly for implantable devices. As medical devices become smaller and more advanced, the need for efficient power consumption becomes more urgent. Implantable devices, such as pacemakers, insulin pumps, and hearing aids, require a constant power supply for operation, but the space available for batteries is limited. This makes it difficult to ensure that the devices will function effectively for long periods without requiring frequent battery replacements or recharging.

At the same time, the antennas used for communication in these devices need to transmit signals over distances and through varying mediums, which often requires a significant amount of energy. The challenge, therefore, lies in balancing the need for high power transmission with the limitations imposed by the device's battery size. Additionally, as medical devices are used over extended periods of time, the challenge is to maintain the power efficiency of the antennas without compromising their communication range or reliability.

One promising solution to this issue lies in the field of energy harvesting. Researchers are exploring ways to integrate energy harvesting mechanisms into biomedical devices, such as using body movement, temperature gradients, or even ambient light to generate power. By capturing and converting ambient energy, devices could potentially extend their operational lifetime without the need for external charging or frequent battery replacements. Furthermore, the development of wireless power transfer (WPT) systems, which enable power to be transferred to devices wirelessly from external sources, could greatly reduce the need for batteries altogether. Energy-efficient antennas, combined with advanced energy harvesting and WPT technologies, would represent a significant breakthrough in the field of biomedical antennas.

10.5.2.2 Biocompatibility and Safety

Another crucial challenge in the development of biomedical antennas is ensuring biocompatibility. When antennas are implanted in or near the human body, they must not cause any adverse biological reactions. The materials used in the construction of these antennas must be safe for long-term contact with human tissues and fluids, and they must not provoke immune system responses or lead to chronic inflammation. Moreover, the antenna must not degrade or corrode over time, which could compromise both its performance and safety.

Research into the biocompatibility of materials has been ongoing for several years, with advancements in the development of biocompatible metals, polymers, and ceramics. Titanium and its alloys, for example, are commonly used in medical implants due to their biocompatibility, strength, and corrosion resistance. Similarly, polymers such as polyimide and polytetrafluoroethylene (PTFE) are being investigated for their flexibility, durability, and ability to withstand harsh bodily conditions.

The challenge, however, is not just selecting the right materials, but also ensuring that the antenna is designed in a way that minimizes any risk of infection or tissue irritation. This requires carefully considering the shape, size, and method of implantation to avoid excessive pressure on surrounding tissues or blood vessels. Furthermore, as antennas are often part of larger, more complex devices, the biocompatibility of the entire device must be considered, including the antenna's interaction with other materials and components.

10.5.2.3 Integration with the Human Body

Successful integration of antennas into the human body remains one of the most significant hurdles in the field of biomedical antenna technology. The human body is a complex, dynamic environment that can affect the performance of medical devices, particularly antennas. The interaction between the antenna and the body can lead to signal degradation, detuning, and other performance issues, especially when antennas are placed near or embedded within the body. Body tissues can absorb or reflect signals, leading to reduced transmission range and communication reliability.

The challenge is to design antennas that can perform effectively and consistently despite these challenges. This requires a deep understanding of the human body's electromagnetic properties and the development of body-centric communication models that account for the varying conditions created by the body. Adaptive antennas that can adjust their characteristics in response to changes in the surrounding environment hold great promise in overcoming these challenges. For example, antennas that can dynamically adjust their radiation pattern or impedance based on the position of the body or surrounding tissues could help mitigate the impact of body-induced interference.

Flexible and stretchable antenna designs also hold promise for improving the integration of antennas with the human body. These antennas can conform to the shape of the body and maintain a close and stable contact with tissues, improving both signal transmission and patient comfort. Flexible antennas can also be incorporated into wearable devices that monitor health metrics in real time, allowing for continuous tracking without the need for bulky or uncomfortable equipment.

10.5.2.4 Cost-Effective Manufacturing

The high cost of manufacturing biomedical antennas remains another significant barrier to their widespread adoption. Advanced materials, precision manufacturing techniques, and miniaturization all contribute to the high cost of these devices. As the demand for biomedical antennas grows, it will be crucial to develop cost-effective manufacturing methods to ensure that these technologies are accessible to a broader range of patients.

One potential solution lies in the development of scalable manufacturing techniques, such as 3D printing and roll-to-roll processing. 3D printing allows for the rapid prototyping and fabrication of complex antenna designs with minimal material waste. This technology has the potential to significantly reduce production costs and time, making it easier to produce customized antennas that meet specific patient needs. Roll-to-roll processing, commonly used in the production of flexible electronics, could also enable high-volume, low-cost production of biomedical antennas, particularly for wearable or disposable devices.

In addition to cost-effective manufacturing, efforts must be made to streamline the regulatory approval process for biomedical antennas. Standardized testing and certification procedures will help ensure that antennas meet safety, performance, and biocompatibility standards, reducing the time and cost associated with bringing new technologies to market.

10.5.3 Looking Ahead: The Next Generation of Biomedical Antennas

The future of biomedical antenna technology holds immense promise, driven by continued advances in materials science, miniaturization, and wireless communication. Researchers are constantly exploring innovative solutions to address the current challenges, and we are likely to see significant breakthroughs in the coming years. The integration of advanced antenna designs with wearable, implantable, and diagnostic devices will help to improve patient care, reduce healthcare costs, and enable more personalized, real-time health monitoring.

However, the realization of this potential will require interdisciplinary collaboration across various fields of research, including electrical engineering, biomedical engineering, material science, and manufacturing. The future of biomedical antenna technology is bright, but overcoming the challenges that remain will require continued innovation, careful design, and rigorous testing to ensure that these devices are efficient, safe, and effective. As research in this area continues to evolve, it is expected that biomedical antennas will play an increasingly vital role in the healthcare sector, providing the infrastructure needed for a new era of personalized and connected healthcare solutions.

References

1. D. Venkatachalam et al., Compact flexible planar antennas for biomedical applications: insight into materials and systems design. Bioengineering **10**(10), 1137 (2023)
2. S. Suresh et al., Application of body sensor networks in health care using the internet of medical things (IoMT), in *The Next Generation Innovation in IoT and Cloud Computing with Applications*, (CRC Press), pp. 131–156
3. A. Akinola, G. Singh, U. Nissanov, Performance analysis of microstrip antenna with several feed networks for satellite communication. Int. J. Commun. Antenna Propag. **11**(5), 304–313 (2021)
4. G. Kaur, A. Kaur, G.K. Toor, et al., Antennas for biomedical applications. Biomed. Eng. Lett. **5**, 203–212 (2015). https://doi.org/10.1007/s13534-015-0193-z
5. A. Kiourti, K.S. Nikita, A review of implantable patch antennas for biomedical telemetry: challenges and solutions [wireless corner]. IEEE Antennas Propag. Mag. **54**(3), 210–228 (2012)
6. S. Hussain et al., Modern diagnostic imaging technique applications and risk factors in the medical field: a review. Biomed. Res. Int. **2022**, 5164970 (2022)
7. A. Kiourti et al., Next-generation healthcare: enabling technologies for emerging bioelectromagnetics applications. IEEE Open J. Antennas Propag. **3**, 363–390 (2022)
8. A. Segun, G. Singh, P. Thukur, M. Sharma, Beam steering antenna design with high gain for satellite communication, in *CNC 2020. Communications in Computer and Information Science*, vol. 2, (Springer, Singapore, 2021), pp. 366–379
9. J.M. Bangeles, E.M. Tapawan, D.E. Arboleda, Wearable antennas in medical applications: a comprehensive review of design, performance, safety, and clinical applications. Int. J. Res. Publ. Rev. **5**, 709–723 (2024)
10. E. Apaydin, *Microfabrication Techniques for Printing on PDMS Elastomers for Antenna and Biomedical Applications*., Ph.D. dissertation, (The Ohio State Univ, 2009)
11. Y. Jiang, T. Liu, Y. Zhou, Recent advances of synthesis, properties, film fabrication methods, modifications of poly(3,4-ethylenedioxythiophene), and applications in solution-processed photovoltaics. Adv. Funct. Mater. **30**(51), 2006213 (2020)
12. S. Dwivedi, Nano-metamaterial devices: Design of Antenna for beyond 5G applications, in *Handbook of Nano-Metamaterials*, (Springer, Singapore, 2024), pp. 1–27
13. R. Kalaiyarasan, R.B. Bibi, G. Nagarajan, An exclusive survey on metamaterial-based fractal antennas for wireless applications, in *Design and Optimization of Wearable, Implantable, and Edible Antennas*, (IGI Global, 2024), pp. 18–32
14. M. Wagih et al., Microwave-enabled wearables: underpinning technologies, integration platforms, and next-generation roadmap. IEEE J. Microwaves. **3**(1), 193–226 (2022)
15. M. Chen, S. Gonzalez, A. Vasilakos, H. Cao, V.C.M. Leung, Body area networks: a survey. Mobile Netw. Appl. **16**, 171–193 (2011)
16. S.N. Mahmood et al., ON-OFF body ultra-wideband (UWB) antenna for wireless body area networks (WBAN): a review. IEEE Access **8**, 150844–150863 (2020)
17. W. Su, J. Zhu, H. Liao, M.M. Tentzeris, Wearable antennas for cross-body communication and human activity recognition. IEEE Access **8**, 58575–58584 (2020)
18. M. Gezimati, G. Singh, Terahertz imaging and sensing for healthcare: current status and future perspectives. IEEE Access **11**, 18590–18619 (2023)

19. Y. Zhou, C. Liu, Y. Huang, Wireless power transfer for implanted medical application: a review. Energies **13**(11), 2837 (2020)
20. K. Agarwal, R. Jegadeesan, Y.-X. Guo, N.V. Thakor, Wireless power transfer strategies for implantable bioelectronics. IEEE Rev. Biomed. Eng. **10**, 136–161 (2017)
21. H. Dinis, I. Colmiais, P.M. Mendes, Extending the limits of wireless power transfer to miniaturized implantable electronic devices. Micromachines **8**(12), 359 (2017)
22. A. Bontempi, *Wireless Power Transfer for Implantable Medical Devices.*, Ph.D. dissertation (Politecnico di Torino, 2021)
23. R. Sedehi et al., A wireless power method for deeply implanted biomedical devices via capacitively coupled conductive power transfer. IEEE Trans. Power Electron. **36**(2), 1870–1882 (2020)
24. A. Segun, U. Nissanov, G. Singh, Design and analysis of dual rectangular slotted antenna for space-station cynosure IOT security applications, in *Proc. RDAAPS*, (2021), pp. 1–8
25. H. Mohammad, R. Shadid, Wireless power transfer approaches for medical implants: a review. Signals **1**(2), 209–229 (2020)
26. S. Rajkumar, Edible antennas: design, development, and, in *RFID, Microwave Circuit, and Wireless Power Transfer Enabling 5/6G Communication*, (2025), p. 367
27. F. Mokhtari et al., Piezofibers to smart textiles: a review on recent advances and future outlook for wearable technology. J Mater Chem A **8**(19), 9496–9522 (2020)
28. A.R. Chishti et al., Advances in antenna-based techniques for detection and monitoring of critical chronic diseases: a comprehensive review. IEEE Access **11**, 104463–104484 (2023)
29. C. Origlia et al., Review of microwave near-field sensing and imaging devices in medical applications. Sensors **24**(14), 4515 (2024)
30. A.I. Kanaan, A.M.A. Sabaawi, *Implantable wireless systems: a review of potentials and challenges* (IntechOpen, 2021)
31. M.A. Hannan et al., Energy harvesting for the implantable biomedical devices: issues and challenges. Biomed. Eng. Online **13**, 79 (2014)
32. S. Roy et al., Powering solutions for biomedical sensors and implants inside the human body: a comprehensive review on energy harvesting units, energy storage, and wireless power transfer techniques. IEEE Trans. Power Electron. **37**(10), 12237–12263 (2022)
33. A.S.Y. Poon, Miniaturized biomedical implantable devices, in *Implantable Bioelectronics*, (2014), pp. 45–64
34. K. Aliqab, I. Nadeem, S.R. Khan, A comprehensive review of in-body biomedical antennas: design, challenges and applications. Micromachines **14**(7), 1472 (2023)
35. M.M. Amri, U.K. Hisan, D. Sulisworo, Implantable antennas for biomedical purposes: state-of-the-art, challenges, and future directions. Indones. Rev. Phys. **7**(1), 1–22 (2024)
36. V.K. Khanna, *Implantable medical electronics*, vol 10 (Springer, 2016)

Index

© The Editor(s) (if applicable) and The Author(s), under exclusive license to Springer Nature Switzerland AG 2026
A. Segun Ayokunle et al., *Biomedical Antenna Technology*,
https://doi.org/10.1007/978-3-031-95268-5

MIX
Papier aus verantwortungsvollen Quellen
Paper from responsible sources
FSC® C105338

If you have any concerns about our products,
you can contact us on
ProductSafety@springernature.com

In case Publisher is established outside the EU,
the EU authorized representative is:
Springer Nature Customer Service Center GmbH
Europaplatz 3, 69115 Heidelberg, Germany

Printed by Libri Plureos GmbH
in Hamburg, Germany